Anatomy and Physiology

Anatomy and Physiology Made Easy: A Concise Learning Guide to Master the Fundamentals

Dr. Phillip Vaughn

ISBN: 1534635319
ISBN-13: 978-1534635319

Contents

Section 1: Introduction

"The human body is the most complex system ever created. The more we learn about it, the more appreciation we have about what a rich system it is."

-Bill Gates

Both human anatomy and physiology are closely related to biology – the study of live and of living organisms. Anatomy is the science of body structures and physiology is the way in which these body structures function and cooperate to support the processes that enable life. In other words, anatomy refers to the structure and physiology to the function of the human body. Physiology is experimental, whereas anatomy is a descriptive discipline, and these of course go hand and hand because structure and function influence each other reciprocally.

This book adopts a layered approach. It begins with an introduction to anatomical terms of location and orientation, before looking at the basic structural and functional unit of the life: the cell. In the fourth section, the structural organization of the human body and essential medical terminology are covered. This will equip you with the fundamental knowledge you need to embark upon your voyage around the human body. What follows is a thematic presentation of the essential body systems of the human body and their structural and functional significance.

The aim of this book is to help you navigate the human body and understand the fascinating structures and processes that determine your biorhythm. It is tailored toward the lifelong learner and explorer and can also be used by students dipping into the subject, along with their anatomy and physiology text.

Lastly, it is important to remember that there is always more to learn. Despite the fact that we have been studying the human body for hundreds of years, scientists make new discoveries all the time. I hope that you enjoy this introduction to anatomy and physiology and that it marks the beginning of a lifelong exploration and appreciation for the human body.

Best wishes,

Dr. Phillip Vaughn

Section 2: The Human Body

Medical terminology is the language scientists and medical practitioners use to avoid ambiguity and effectively communicate with one another in an accurate and science-based manner. To describe human anatomy and physiology, medical terminology is used to combine several words into one word, thereby making communication significantly more efficient and precise. Because anatomy is a descriptive discipline, concerned with shape and structure, understanding medical terminology and using anatomic terms is crucial to navigate the human body.

In this section, we will look at the anatomic terms that are used to describe directions within the human body, as well as those referring to body planes, cavities, and regions.

1. Directional Terms

Directional terms are used to determine the exact location of a structure within the body. Directional root words are the foundation of directional terms that refer to a certain part or area of the body.

Directional Root Word = Meaning

Anter(o) = Front

Caud(o) = Tail, downward

Cephal(o) = Head, upward

Dist(o)= = Away from

Dors(o) = Back

Infer(o) = Below

Later(o) = Side

Medi(o) = Middle

Poster(o) = Back, behind

Proxim(o) = Near (proximate)

Super(o) = Above

Ventr(o) = Front, belly

Anterior and **ventr**al are used when referring to the front of the body. The kneecap for example, is found on the anterior side of the leg.

Posterior and **dors**al are used when referring to the back of the body. The shoulder blades for example, are located on the posterior side of the body.

Cephalad and **super**ior are used when referring to 'above the waistline.' The hand for example, is part of the superior extremity.

Caudal and **infer**ior are used when referring to 'below the waistline'. The foot for example, is part of the inferior extremity.

Lateral is used when referring to the sides of the body. The little toe for example, is found at the lateral side of the foot.

Medial is used when referring to the middle of the body. The middle toe for example, is found at the medial side of the foot.

Proximal ("closest") is used when referring to the center of the body or of the point of attachment. The proximal end of the femur for example, joins with the pelvic bone.

Distal ("farthest") is used when referring to the outer part of the body, away from the point of attachment or origin. The hand for example, is located at the distal end of the forearm.

Superficial in its medical meaning refer to something that is 'on the surface' or 'shallow'. The skin for example, is superficial to the muscles. Furthermore, the cornea is to be found on the superficial surface of the eye.

Intermediate ("between") is used to refer to a structure being or occurring at the middle place. The abdominal muscles for example, are intermediate between the small intestines and the skin.

Deep refers to 'farther away from the surface.' The abdominal muscles for example, are deep to the skin.

Unilateral means that a structure is located/to be found 'on only one side of the body', such as the liver and the stomach.

Bilateral means that a structure is located/to be found 'on both sides of the body', such as the arms, legs, eyes, and kidneys.

Ipsilateral means that a structure is located/to be found 'on the same side of the body'. For example, it could be said that the right ear and the right eye are ipsilateral to one another.

Contralateral means that a structure is located/to be found 'on the opposite side of the body'. The left ear for example, is contralateral to the right ear.

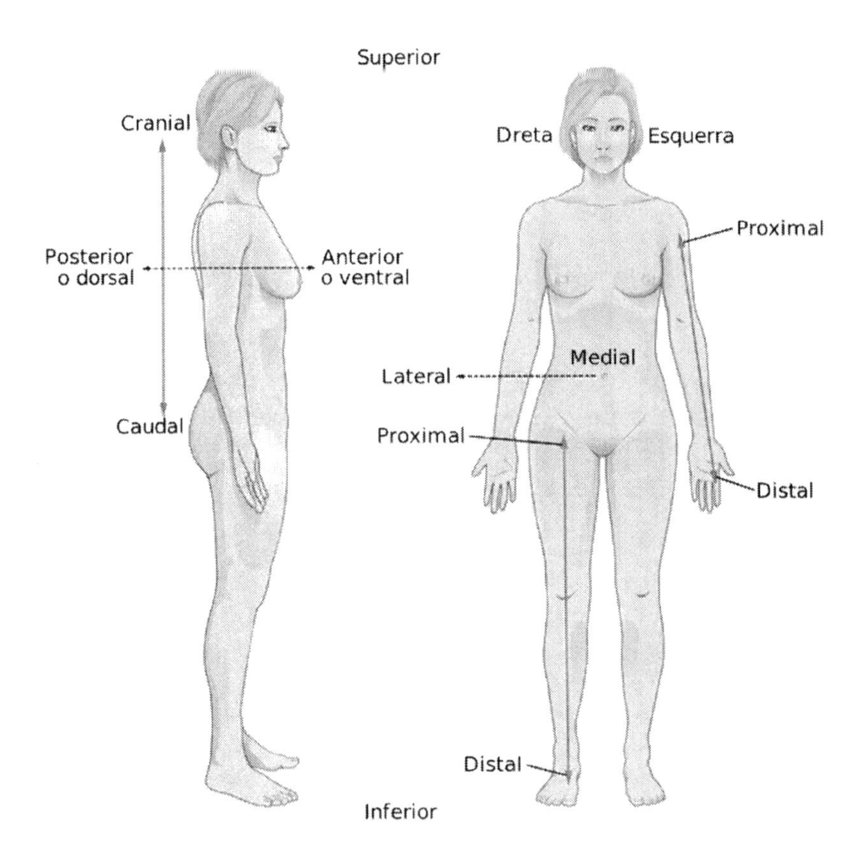

2. Body planes

Body planes or anatomical planes are hypothetical lines used identify a specific location or area by transecting the human body into different parts.

In human anatomy, three basic anatomical planes are used to divide the body from right to left, back to front, and top from bottom.

- **The Frontal or Coronal Plane**: This is a vertical plan that separates the front from the back of the body, i.e. the anterior form the posterior and the ventral from the dorsal.
- **The Sagittal Plane**: The sagittal plane, also known as lateral, is a vertical plan that separates the body into right and left sides. **The mid-sagittal or median plane** is a specific sagittal plane that divides the body into right and left at the body's exact midline.
- **The Transverse Plane:** The transverse plane, also known as horizontal or axial plane, is a horizontal plane that is runs parallel to the ground and through the waistline. It divides the body into upper and lower halves.

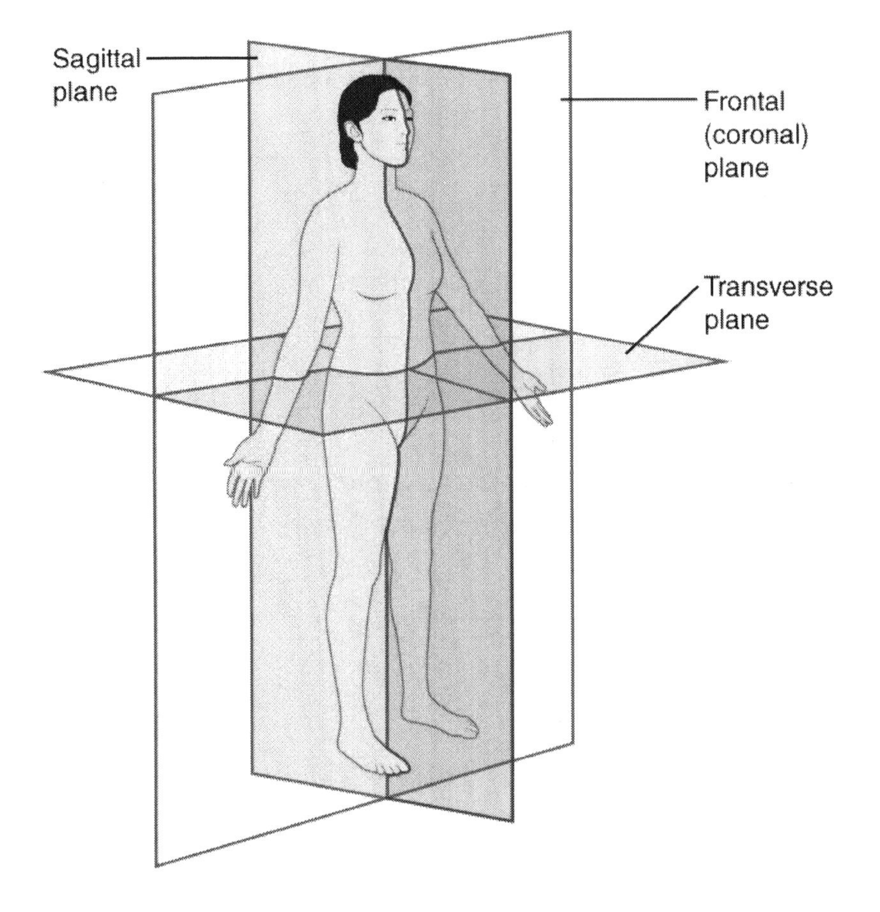

3. Cavities

Body cavities are fluid-filled spaces that contain the internal, visceral or splanchnic body organs. The dorsal body cavity and the ventral body cavity are the two major closed cavities of the human body.

The Dorsal Body Cavity, located on the posterior region of the body, comprises of:

- **The Cranial Cavity**, which is enclosed by the skull and contains the brain. The cranial cavity is also called the calvaria.
- **The Spinal Cavity,** which contains the spinal cord. The spinal cavity is also called the vertebral cavity or vertebral canal.

The Ventral Body Cavity, located on the anterior region of the body, comprises of:

- **The Thoracic or Chest Cavity:** This cavity contains the esophagus, lungs, trachea, heart, and aorta.
- **The Abdominopelvic Cavity**, which comprises of:
    - **The Abdominal Cavity:** This cavity contains the stomach, intestines, liver, spleen, gallbladder, pancreases, ureters, and kidneys.
    - **The Pelvic Cavity:** This cavity contains the urinary bladder, urethra, rectum, uterus, part of the large intestine, and the reproductive organs.

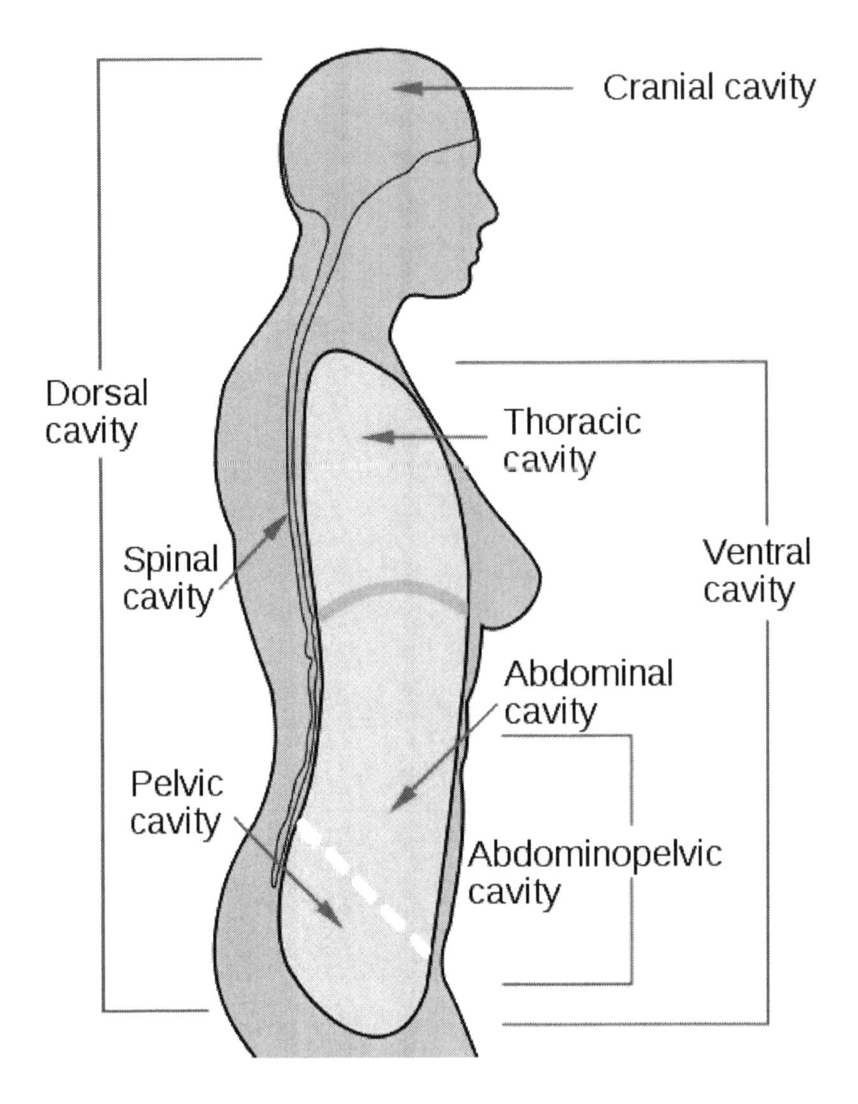

Besides the two major cavities, the body also contains smaller cavities:

- **The nasal cavity** (nose)
- **The oral cavity** (mouth)
- **The orbital cavities** (eyes)
- **The synovial cavities** (joint cavity)

- **The tympanic cavities** (a small cavity which surrounds the bones of the middle ear, also called middle ear cavities)

4. Regions

Body regions are areas of the body that perform special functions or are supplied by specific blood or nerve cells. In this section, we will cover the most widely used body region terms - those belonging to the abdominal area - as well as other important small regions of the body.

THE ABDOMINAL AREA

The abdominal area is divided into nine anatomic regions, the division of which facilitates the diagnosis of abdominal problems.

The portion in the center is the **umbilical region**. This is the area that surrounds the umbilicus (navel). The organs that are found in the umbilical region include the Umbilicus, Jejunum, Ileum, and the Duodenum. On either side of the umbilical region are the right and left lumbar regions. The **right lumber region** comprises of the Gallbladder, the Liver, and the Right Colon. The left lumber region comprises of the Descending Colon and the Left Kidney.

The region directly above the umbilical region is called the **epigastric region.** The organs that are found in this region include the Stomach, Liver, Pancreas, Duodenum, Spleen, and the Adrenal Glands. To the right and left of the epigastric region are the right and left hypochondriac regions. The **right hypochondriac region** comprises of the Liver, Gallbladder, Right Kidney, and the Small Intestine. The **left hypochondriac region** comprises of the Spleen, Colon, Left Kidney, and the Pancreas.

The region directly below the umbilical region is called the **hypogastric region**, which is the region of the abdomen below the navel. The organs that are found in the hypogastric region include the following: Urinary Bladder, Sigmoid Colon, and the Female Reproductive Organs. To the right and left of the hypogastric region are the right and left iliac regions. The **right iliac region (or fossa)** comprises of the Appendix and the Cecum. The **left iliac region (fossa)** comprises of the Descending Colon and the Sigmoid Colon,

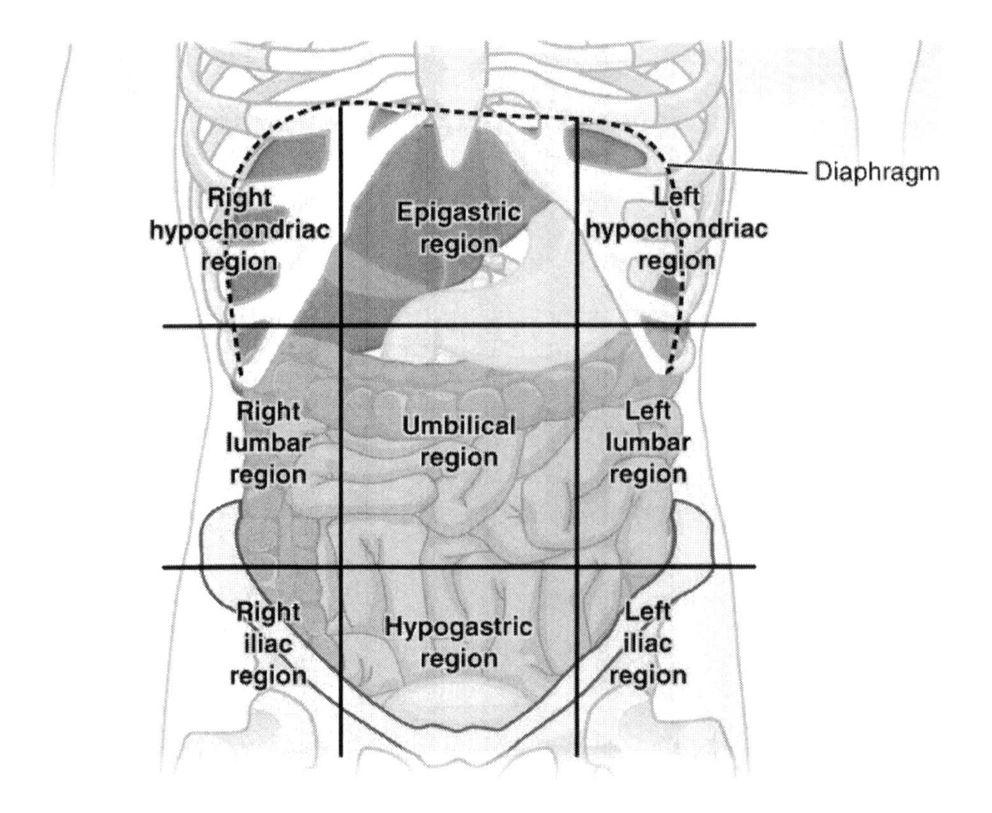

When a patient is being examined, clinical regions are used to divide the abdominal area into four equal quadrants.

- **The Right Upper Quadrant (RUQ)**: This region is often assessed to localize pain as well as tenderness. It is often tender in clients with cholecystitis, hepatitis, or a developing peptic ulcer. Important organs that are found in this quadrant include the liver, the gall bladder, and parts of the small and large intestines.

- **The Left Upper Quadrant (LUQ):** This region is often tender in clients with abnormalities of the intestines and in clients with appendicitis. Important organs that are found in this quadrant include the stomach, pancreas, spleen, the left portion of the liver, and parts of the small and large intestines.

- **The Right Lower Quadrant (RLQ)**: This region, which stretches from the median plane to the right side of the body and from the umbilical plane to the right inguinal ligament, is often tender and painful in clients with appendicitis. Important organs that are found in this quadrant include the appendix, the upper portion of the colon, the right ovary, the right ureter, the Fallopian tube, and parts of the small and large intestines.

- **The Left Lower Quadrant (LLQ):** This region, which is located below the umbilicus plane, is usually tender and painful in clients with ovarian cysts or a pelvic inflammation. Abdominal pain in this region can also be a symptom of colitis, diverticulitis, or ureteral colic. Tumors in the LLQ can be indicative of colon or ovarian cancer. Important organs found in this region include the left ovary, the left ureter, the Fallopian tube, and parts of the small and large intestine.

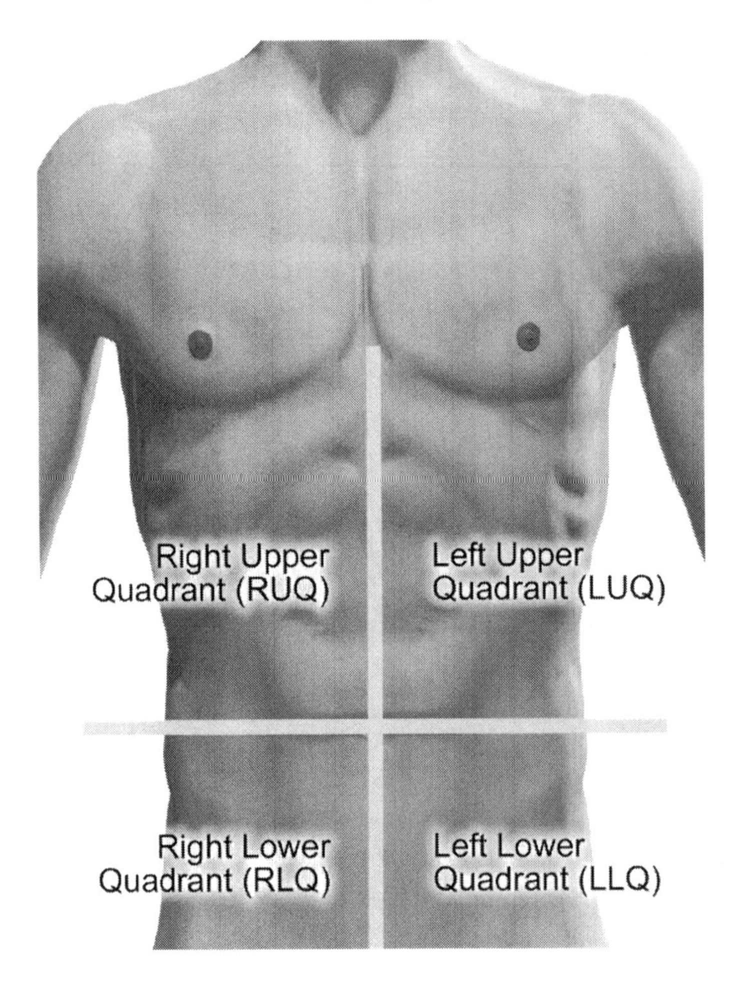

THE SPINAL COLUMN

The spinal column is divided into five regions:

- **The Cervical Region (Abbreviation C)** consists of seven cervical vertebrae, C1 to C7, which are the smallest of the true vertebrae. They are located in the neck region closest to the skill.

- **The Thoracic or Dorsal Region (Abbreviation T or D)** consists of 12 thoracic or dorsal vertebrae: T1 to T12 or D1 to D12. They are located in the chest region of the spine.

- **The Lumbar Region (Abbreviation L),** sometimes referred to as the lower spine, consists of five lumbar vertebrae, L1 to L5. The lumbar region is found at the flank or loin area between the ribs and the hip bone.

- **The Sacral Region (Abbreviation S)**, consists of five bones, S1 to S5, which are located at the bottom of the spine. These five bones are fused together to form one bone: **the sacrum.**

- **The Coccygeal Region**, also referred to as the tailbone, is a small bone at the very bottom of the spine. It is composed of four vertebrae that are fused together to form one bone: **the coccyx.**

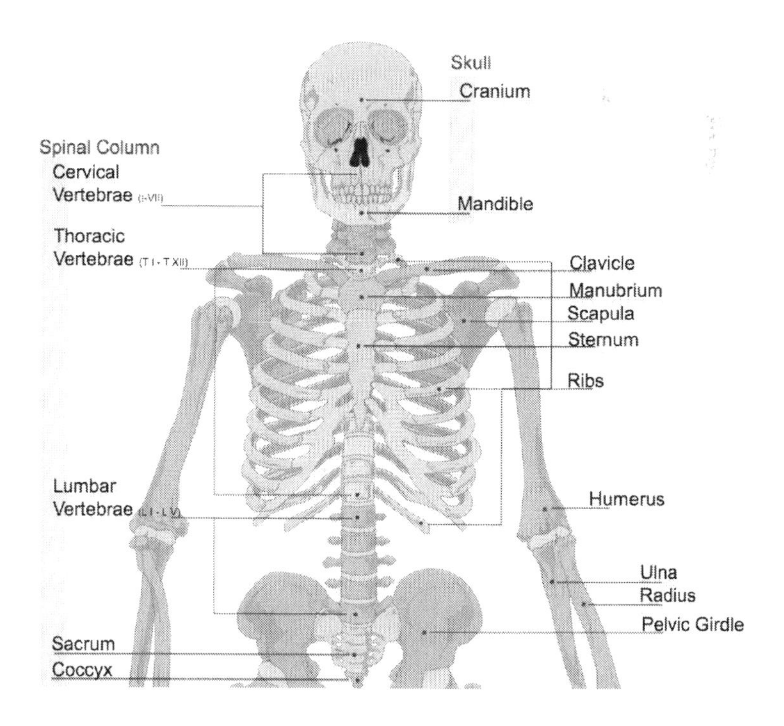

SMALL REGIONS OF THE BODY

- **Auricular Region:** This is the region around the ears.
- **Axillary:** The armpit region.
- **Clavicular:** The region on either side of two slender bones.
- **Infraorbital:** The region below the eyes.
- **Infrascapular:** This is the region on each side of the chest, down to the last rib.
- **Inguinal:** The region of the groin, i.e. the depressed area of the abdominal wall near the thigh.
- **Interscapular:** This is the region on the back between the shoulder blades (the scapulae).
- **Lumbar:** The region of the lower back between the ribs and the pelvis below the infrascapular area.
- **Mammary:** The breast area.
- **Mental:** The chin area.
- **Occipital:** The lower posterior region of the head.
- **Orbital:** The region around the eyes.
- **Pectoral:** The chest area.
- **Perineal:** The region between the anus and the external reproductive organs (the perineum).
- **Popliteal:** The area behind the knee.
- **Pubic:** The area above the pubis and below the hypogastric region.
- **Sacral:** The area above the sacrum, between the hipbones.
- **Sternal:** The area over the sternum.
- **Submental:** The area below the chin.
- **Supraclavicular:** The area above the clavicles.

Section 3: Cells, DNA and Tissues

Now that we have covered the basic medical terminology that you need to navigate the human body, it is time to look into the most basic unit of living organisms: the cell. In this section we will look at the structures of cells, cellular transport, the components and structures of cells, cell reproduction, DNA, RNA, and the types of tissue that the human body contains.

1. Basic Cell Structure

The cell is the smallest functional and structural unit of an organism and the basic unit of all living organisms (excluding viruses). In advanced organisms, including *homo sapiens*, the cell consists of three essential parts: the cytoplasm, bounding membrane, and nucleus.

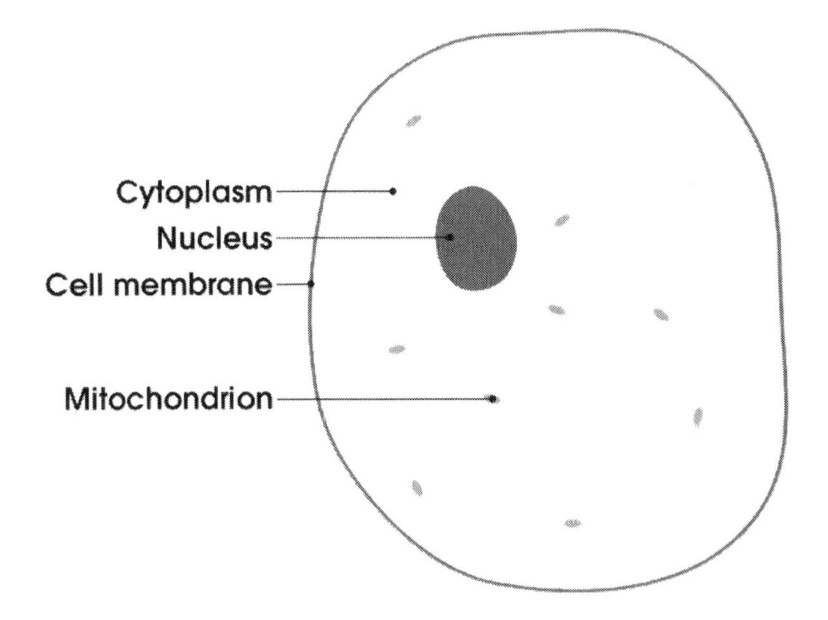

1. **The Cytoplasm**

The cytoplasm is the protoplasm of the body of the cell which surrounds the nucleus. It contains cytosol, organelles, inclusions, and most of the cell's ribonucleic acid (RNA). The protoplasm is the primary component of all cells – both animal and plant cells. Visually, the protoplasm comprises a gel-like matrix that is translucent and viscous in consistency and which contains water, electrolytes (i.e. inorganic ions such as calcium, magnesium, potassium, and sodium), organic compounds (e.g. carbohydrates, lipids, and proteins), and other small molecules.

2. **The Bounding Membrane (Cell Membrane, Plasma membrane or Plasmalemma)**

The cell membrane is the external boundary of a cell that separates it from extracellular fluid and other cells. In all eukaryotic cells, the bounding membrane is made of phospholipid molecules. These molecules form what is known as a phospholipid bilayer, the chemical properties of which allow it to control the substances can and cannot pass through the membrane. In other words, the membrane is semipermeable.

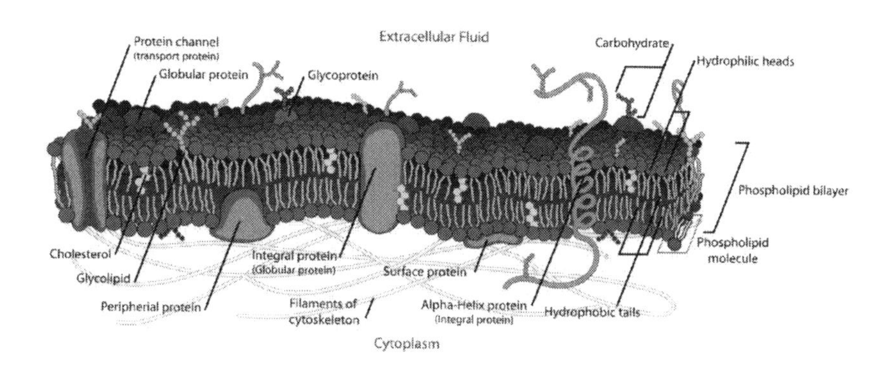

3. The Nucleus

The nucleus is the largest organelle inside the cell, and it directs the cell's activity. It also plays a vital role in cell growth, reproduction, and metabolism. The nucleus contains chromosomes and may also contain one or more nucleoli. Nucleolus (plural nucleoli) is the structure that synthesizes ribonucleic acid (RNA). In eukaryotic cells, chromosomes occur as threadlike strands in the nucleus. Chromosomes contain genetic information, control cellular activity, and are also involved in direct protein synthesis through ribosomes in the cytoplasm. (The structure and function of organelles will be discussed in further depth in ***subsection*** 3).

The nucleus is surrounded by a nuclear envelope that is similar to the cell membrane of the cell. The nuclear envelope contains pores that allow RNA and proteins to pass out of the nucleus, whilst keeping all of the chromatin and nucleolus inside of the nucleus.

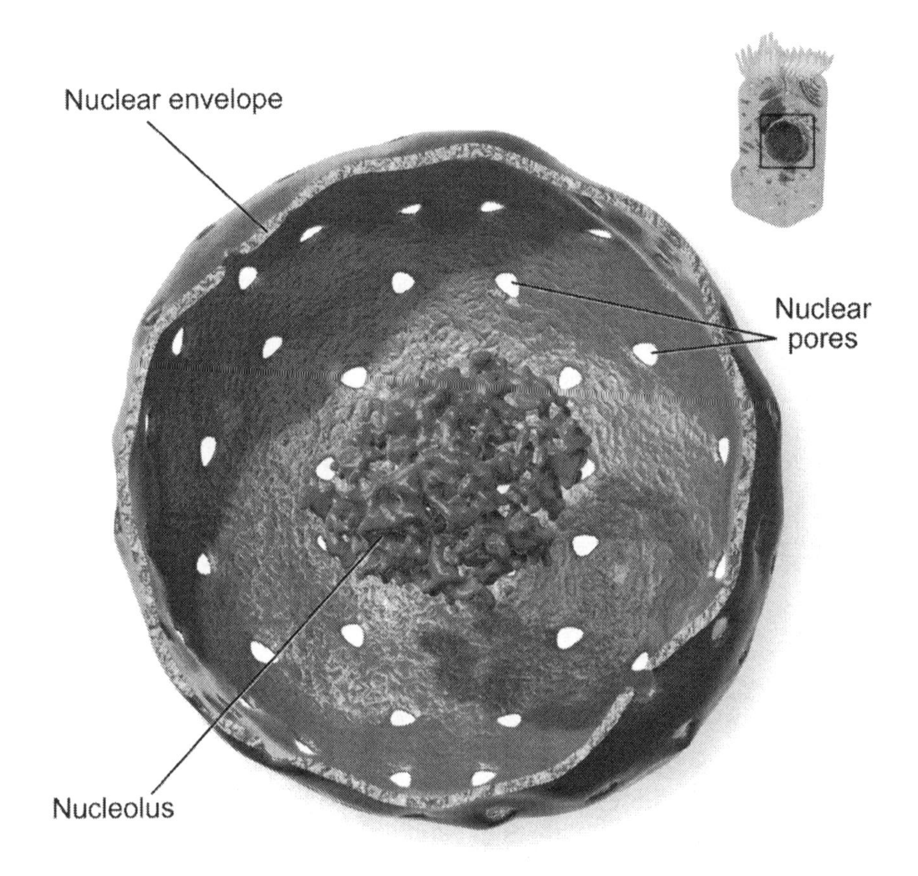
Nuclear envelope
Nuclear
pores
Nucleolus

2. Cellular Transport

Cells interact with other body fluids through the interchange of substances. There are different methods of transportation, some of which are active whilst others are passive in nature. There are three methods by which substances can cross the membrane passively: diffusion, osmosis, and filtration. *Passive transport* means that the substance that passes through the membrane (mostly ions or small molecules) can do so unrestricted, without the need of energy input. *Active transport* is the transport of substances across the cell membrane, which uses energy and requires the assistance of carrier proteins. There are several transport methods - including ***diffusion, osmosis, active transport, endocytosis*** and ***filtration*** - all of which will be covered in this section.

1. Diffusion

In diffusion, substances that are lipid soluble move down a concentration gradient. They move from an area of higher concentration to an area of lower concentration. Lipid soluble substances can *diffuse* through the cell membrane because lipid dissolves in lipid. Movement by way of diffusion continues until concentration distribution is uniform. The rate of diffusion is influenced by a variety of factors including the particle size, the concentration gradient, and the lipid solubility. Generally speaking, the smaller the particles, the greater the rate of diffusion. The greater the concentration gradient, the faster the rate of diffusion. Finally, the more lipid-soluble the particles are, the faster they diffuse through the plasma membrane's lipid layers.

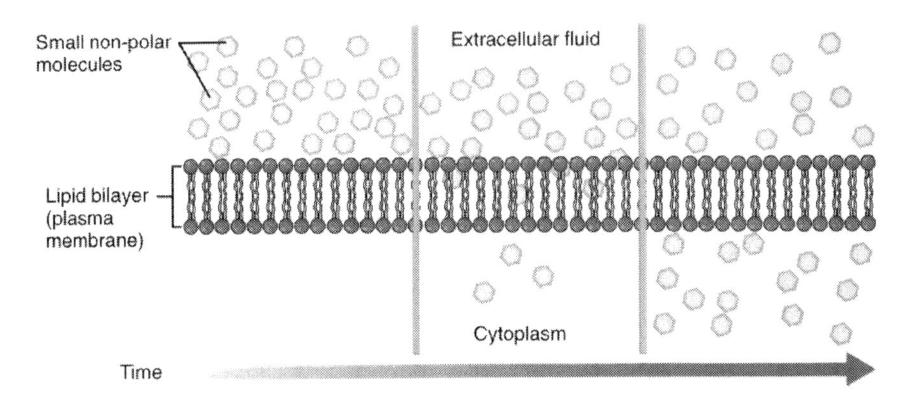

The above diagram illustrates simple diffusion across the plasma membrane. **Diffusion is the movement of lipid soluble substances from an area of higher to an area of lower concentration until an equilibrium is reached.**

Diffusion happens with lipid soluble substances because these can move across the phospholipid cell membrane. Examples of substances which are lipid soluble include oxygen, carbon dioxide, fatty acids, alcohols, and lipid soluble medications. **Non lipid soluble substances move through protein channels in the cell membrane in a process called 'facilitated diffusion'.**

2. Facilitated Diffusion

Other molecules, such as glucose (see diagram below), also move from an area of high concentration to an area of low concentration but do so through a protein channel. Cell membranes contains proteins, some of which span the entire width of the membrane and act as channels for specific substances. These care called *channel proteins* or *carrier proteins* because they essentially 'carry' these non-lipid soluble substances across the plasma membrane by facilitated diffusion.

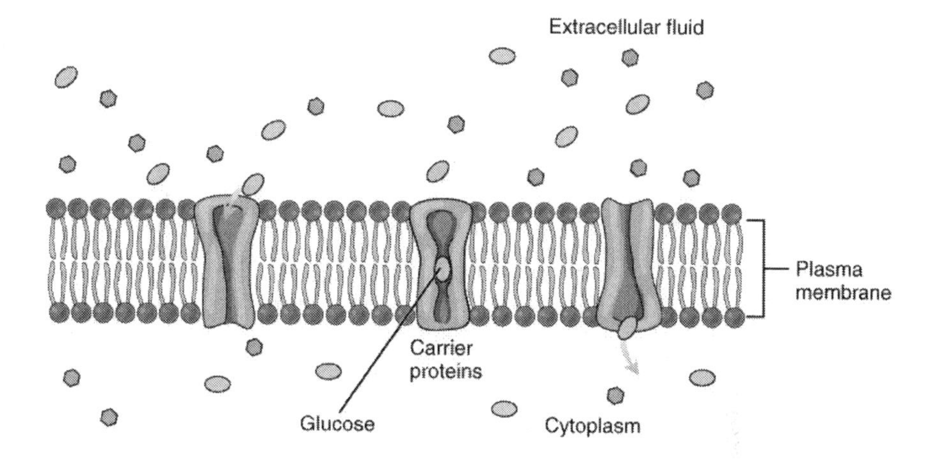

Facilitated diffusion is similar to diffusion because substances move from an area of higher concentration to an area of lower concentration. However, it is distinguishable due to the fact that the substances move through and are *facilitated* by a protein channel. In both cases, movement continues until distribution is uniform.

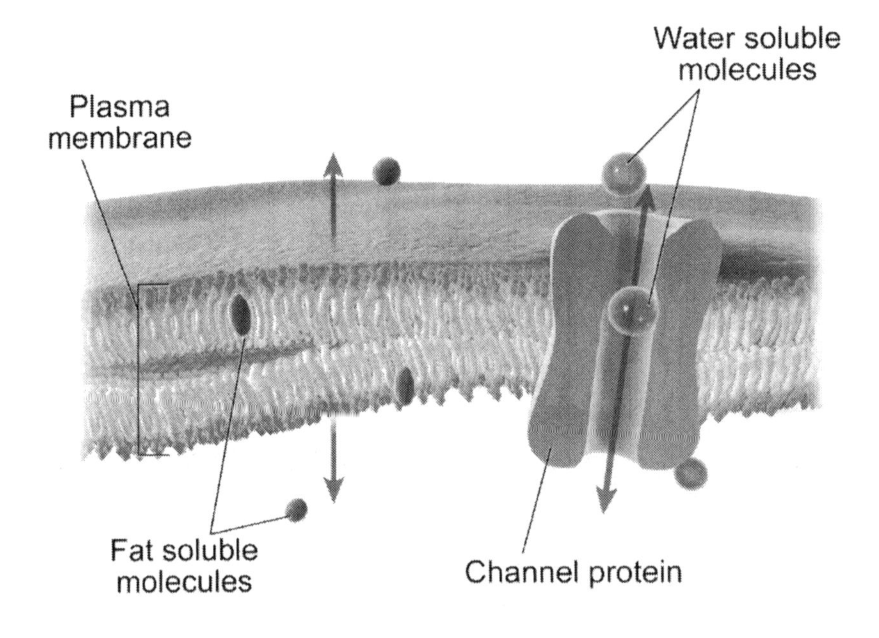

Examples of substances that move by facilitated diffusion include glucose, sodium, potassium, and chloride. The forces that influence the rate of diffusion in facilitated diffusion are the similar to those found in diffusion; a main one being concentration gradient –the greater the concentration gradient or the difference in concentration, the greater the rate of diffusion. In facilitated diffusion however, the rate also depends on the saturation of the receptors on the carrier protein.

As substances move in and out of cells via facilitated diffusion, they must bind to the receptor on the carrier protein (see diagram below). Once the insoluble substance and the carrier protein bind, the protein changes its shape, allowing the substance to pass through the plasma membrane. But because proteins only have a certain amount of receptors, a higher concentration gradient will in some cases not move substances across the membrane at a faster rate. In this case, the rate of facilitated diffusion will depend upon the saturation of receptors on the carrier protein.

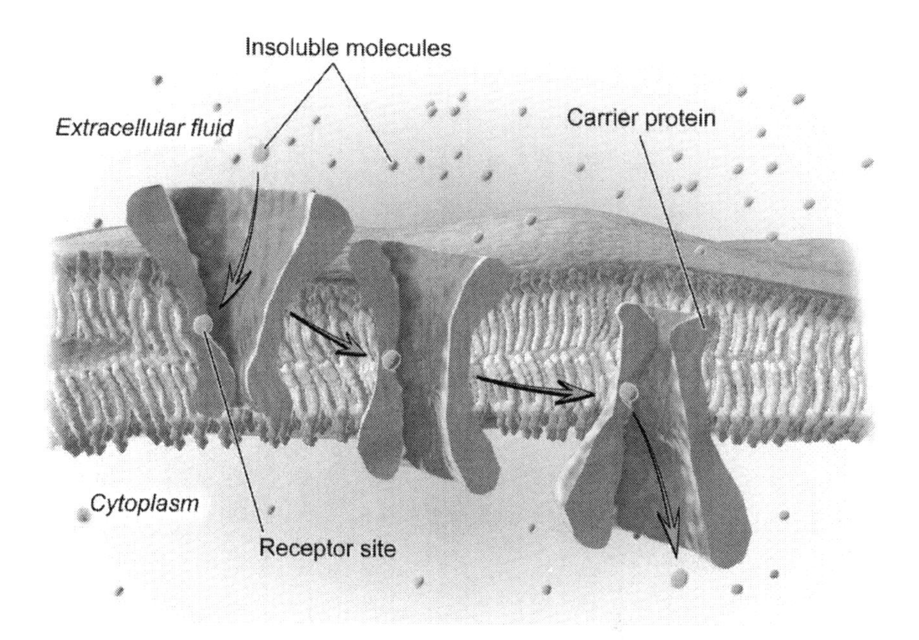

3. Osmosis

The passive transport or movement of water across the semipermeable cell membrane is called *osmosis*. In osmosis, water moves through the semipermeable membrane from an area of lower solute concentration to an area of high solute concentration. Fluid movement continues until the solute concentration on both sides of the membrane equalize.

Fluid moves down the *osmotic gradient*, which is the difference in concentration between two solutions on either side of the cell membrane. The osmotic gradient is generally used when comparing two solutions that are separated by a semipermeable membrane that allows water to move toward the hypertonic solution (i.e. the solution with the higher concentration). An easy way to remember the movement of water in osmosis is that generally "**water follows sodium**" - which means that water always moves toward the area of higher solute concentration.

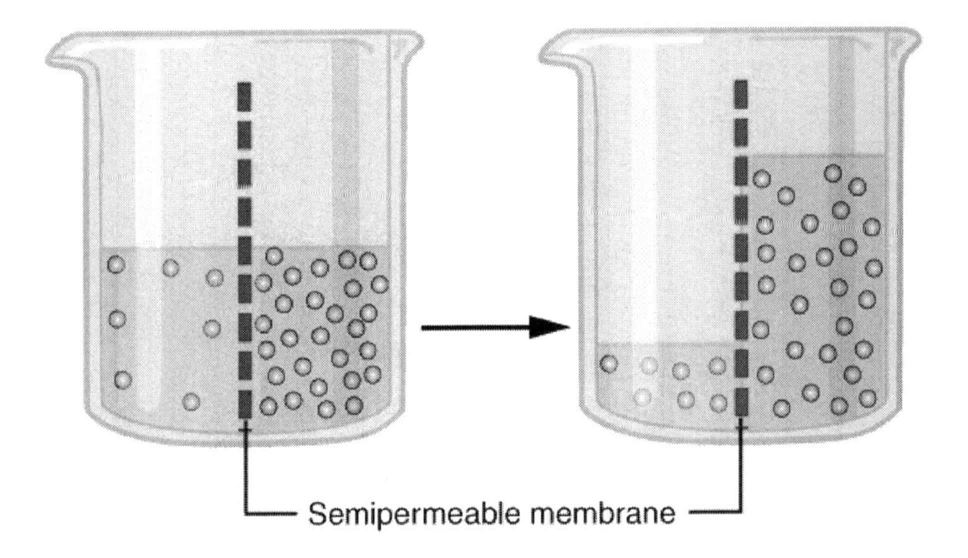

Osmolarity or osmotic concentration (Osm/L or osmol/L) is the measure of a solution's particle concentration, which is defined as the number of osmoles (Osm or osmol) of a solution per litre (L) of a solution. In physiology however, the term tonicity is used when referring to the concentration of solutions across the semipermeable membrane of cells. Tonicity is the relative concentration of solutions that determines not on only the direction of the movement of water, but also its extent or rate. There are three types of tonicity that characterize solute concentration: hypertonic, isotonic, and hypotonic.

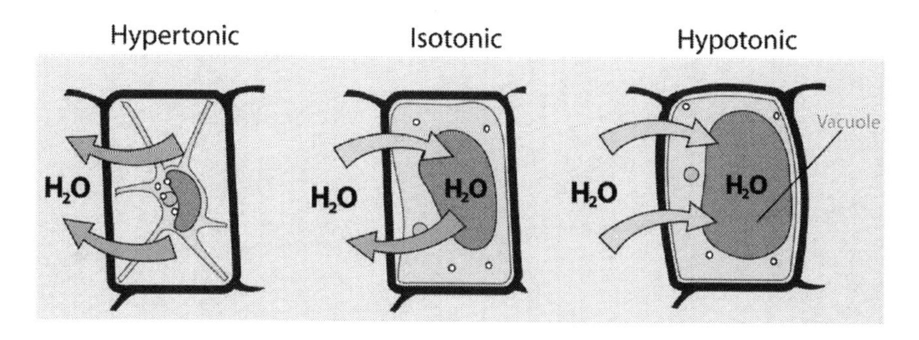

A '**hypertonic**' solution is one that contains a higher concentration of solutes outside the cell than inside the cell. If a solution is more 'concentrated' than the body fluids, that solution is said to be hypertonic. Therefore, if a cell is placed into a hypertonic solution, water will flow out of the cell through osmosis in order to balance out the concentration of solutes between the two solutions (on both sides of the semipermeable membrane). **The cell will thus loose water by osmosis.**

Hypertonic Solution
(Osmotic Flow out of Cell)

The term '**isotonic**' denotes a solution that has the same osmotic pressure as the solution on the other side of the semipermeable membrane. In these cases, as illustrated by diagram below, there is no net movement of water across the cell membrane.

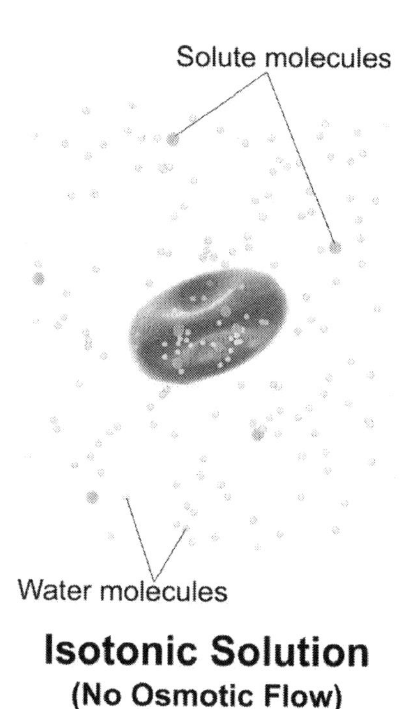

A '**hypotonic**' solution is one that contains a lower concentration of solutes. In other words, if a solution is less concentrated than the body fluids (the concentration of solutes outside the cell are lower than the concentration inside the cell), that solution is said to be hypotonic. If a cell is placed in a hypotonic solution therefore, **the cell will gain water through osmosis.**

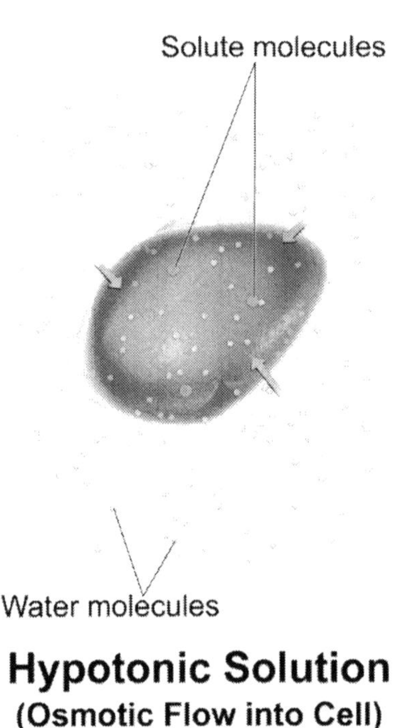

Hypotonic Solution
(Osmotic Flow into Cell)

The effect of osmosis on cells can be illustrated by the following diagram. When red blood cells are placed in a hypertonic solution (as in the left image), they will loose water by osmosis which will cause the cell to crenate (i.e. shrink and acquire a notched appearance). If red blood cells are placed in an isotonic solution (as in the middle image), there will be no net movement of water since the tonicity is equal. Lastly, if the red blood cells are placed in a hypotonic solution, they will gain water by osmosis, causing the cells to swell and potentially burst.

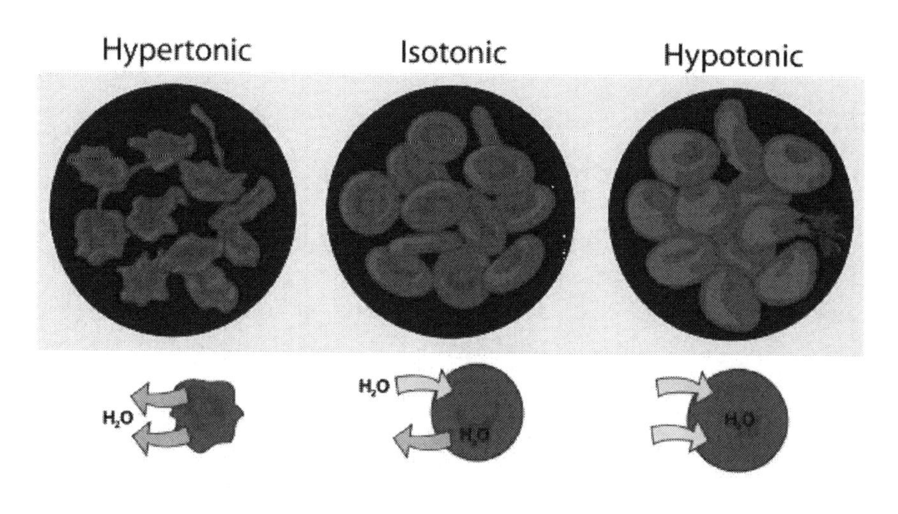

4. Active Transport

Active transport is active because it requires energy. Active transport generally involves the movement or transport of substances **against** the concentration gradient, from an area of low concentration to one of higher concentration, using carrier proteins. The energy that is required to transport substances against their concentration gradients is sourced from ATP. ATP is stored in all cells, providing the energy needed to move molecules and ions in and out of cells.

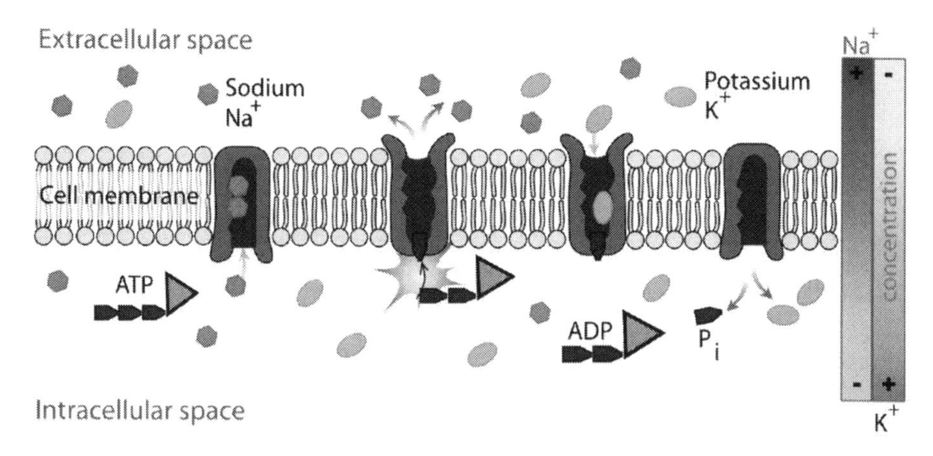

How ATP Generates Cellular Energy

ATP is the 'fuel' which powers processes inside the cell that require energy. Because of this, ATP is sometimes referred to as the "molecular unit of currency". ATP, which stands for adenosine triphosphate, is made up of an adenosine that is joined to three phosphate groups (called a triphosphate). High-energy bonds are found in the chemical bonds between the first and the second phosphate groups, and the second and third phosphate groups. When one phosphate group is removed in a process called **hydrolysis**, energy is released and ATP is converted to ADP (adenosine diphosphate).

Sodium-Potassium Exchange Pump

One example of active an active transport protein is the *sodium potassium pump*. The sodium potassium pump is vital to maintain the workings of the human body and in establishing the concentration gradients essential for life. In the sodium-potassium exchange pump (see diagram below), sodium is transported out of the cell against its concentration gradient, while potassium is transported into the cell against its concentration gradient. The carrier protein uses energy in the form of ATP - one ATP transports three sodium ions out of the cell and two potassium ions inside the cell.

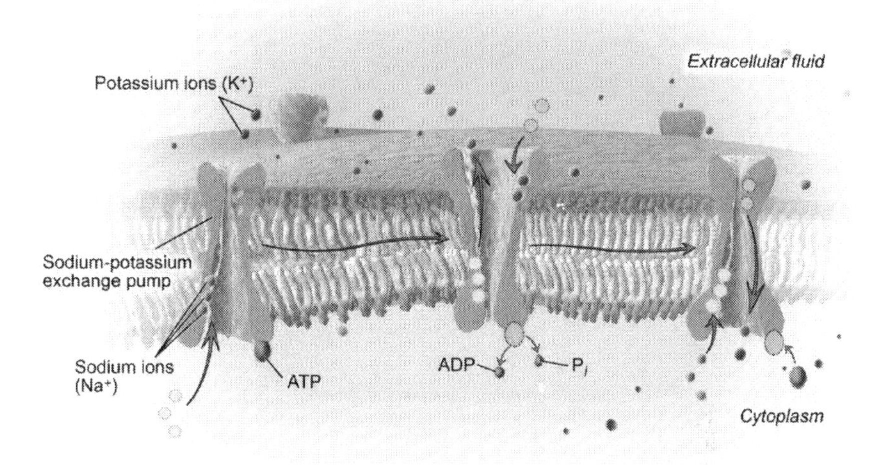

Endocytosis

Endocytosis is a form of active transport in which instead of passing through the cell membrane, the substance is engulfed by the cell itself. The cell engulfs the substance, forming a cavity, called a **vacuole**, which transports the substance into the cell.

Endocytosis involves either *phagocytosis* or *pinocytosis*. Phagocytosis involves the engulfment and ingestion of substances that are too large to pass through the plasma membrane. In phagocytosis, the cell brings a particle into the cell for destruction (ingestion). Pinocytosis occurs in order to bring in small particles or dissolved substances contained in fluid. Receptor-mediated endocytosis is when substances attack to receptors on the plasma membrane.

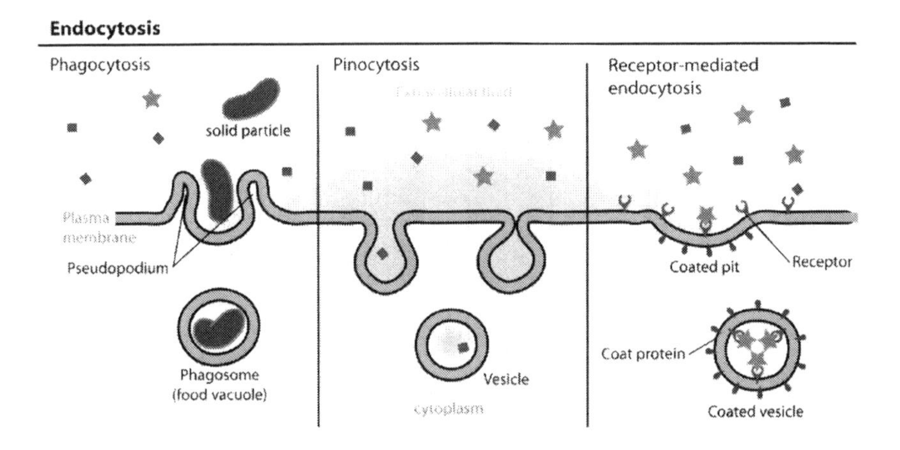

Active transport is the transport of substances that requires the cell to use energy in the form of ATP. Both the sodium-potassium pump and endocytosis are examples of active transport mechanisms.

5. **Filtration**
6.

Fluids and dissolved substances can also move across the cell membrane through **filtration.** In filtration, fluids and other dissolved substances are transported across the capillaries into the interstitial fluid, also called tissue fluid.

The force that drives filtration is called fluid pressure, also called **hydrostatic pressure.** This pressure forces fluids and dissolved particles through the cell membrane. The *rate of filtration,* i.e. how quickly the fluids and dissolved particles pass through the membrane, depends on the amount of pressure that is exerted. Hydrostatic pressure is generated by the cardiovascular system.

3. The Components and Structures of a Cell

Although cells significantly vary in appearance, they are very similar to one another when looking at their inside components. This is particularly true for eukaryotic cells. Eukaryotic cells make up eukaryotes - which includes animals, plants and fungi. Despite the huge variety of different cells, all eukaryotic cells all share fundamental components and structures as outlined in the cross section below. Each single component plays an important role in upholding the health of the cell.

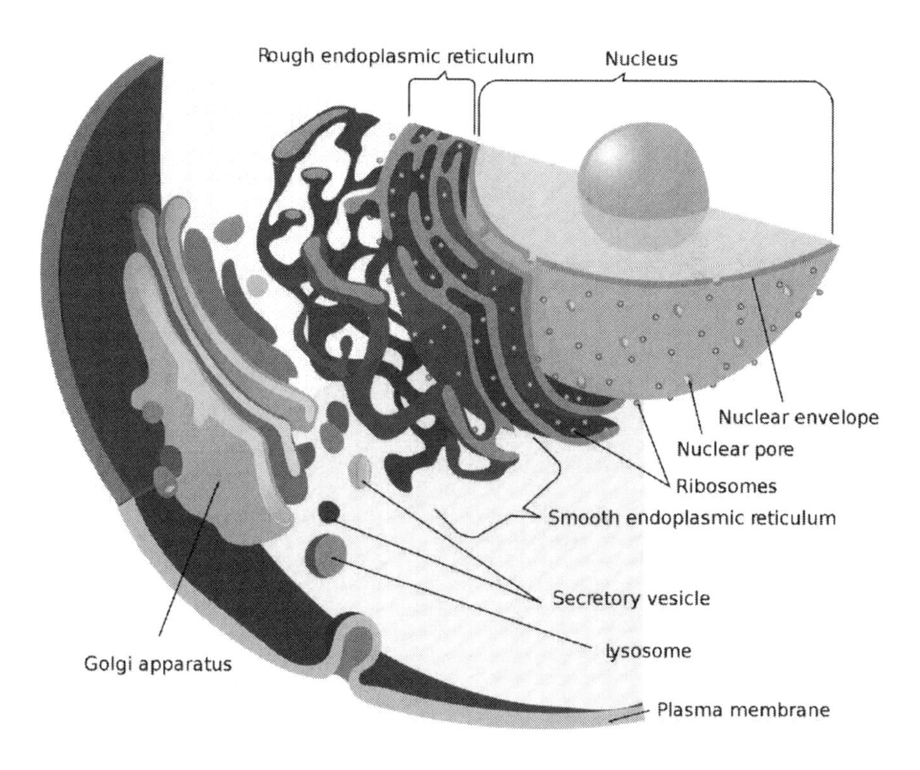

The cross section shows the general structure of a eukaryotic cell, the cell being a membrane-bound structure that contains organelles (the functional subunits of a cell) that are hosted within the cytoplasm.

Organelles (*little organs*) are the metabolic units of a cell. The nucleus (one of the largest organelles) controls the functioning of the cell. Like the nucleus, each organelle performs a specific function that maintains the life of a cell.

ORGANELLE	FUNCTION
Centrioles	Involved in cell division (specifically in the development of spindle fibers). Centrioles occur in pairs and are found near the nucleus.
Endoplasmic Reticulum	Transports protein and lipid components and plays a crucial role in protein synthesis. The **rough endoplasmic reticulum** produces certain proteins and is covered with ribosomes. The **smooth endoplasmic reticulum** contains enzymes that synthesize lipids.
Golgi Complex/Golgi Apparatus	Involved in secretion and intracellular transport. 1. Processes and

	packages protein in sacs called vesicles. Vesicles are composed of a lipid bilayer and are used to transport materials from one place to another. 2. Each apparatus also synthesizes carbohydrate molecules that combine with proteins (produced by the rough endoplasmic reticulum) to form secretory products (e.g. lipoproteins).
Lysosome	Contain digestive enzymes that break down waste material (foreign or damaged material) and harmful cell products and transport them out of the cell via active transport.
Mitochondrion	This is the cell 'powerhouse', i.e. the production site from where cellular energy is generated (in the form of adenosine triphosphate).
Nucleus	Houses genetic material and controls the cell.
Ribosomes	Sites of protein synthesis.

Vacuoles	Vacuoles are membrane-bound spaces (enclosed compartments) in the cytoplasm of the cell. They usually contain water and small molecules and are sometimes also involved in active transport.

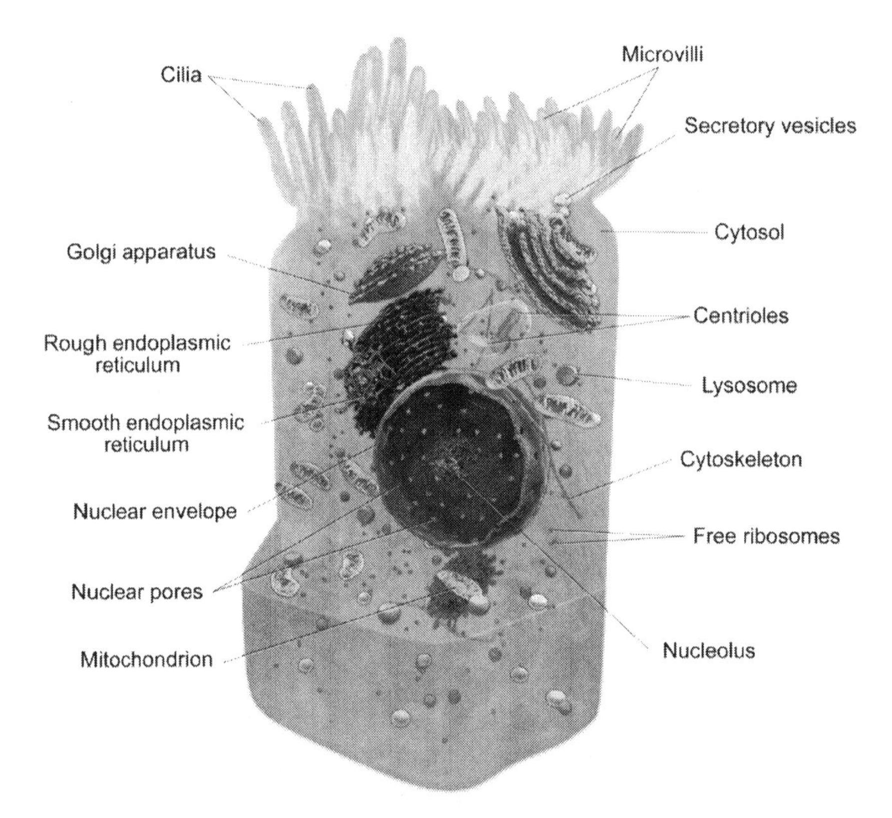

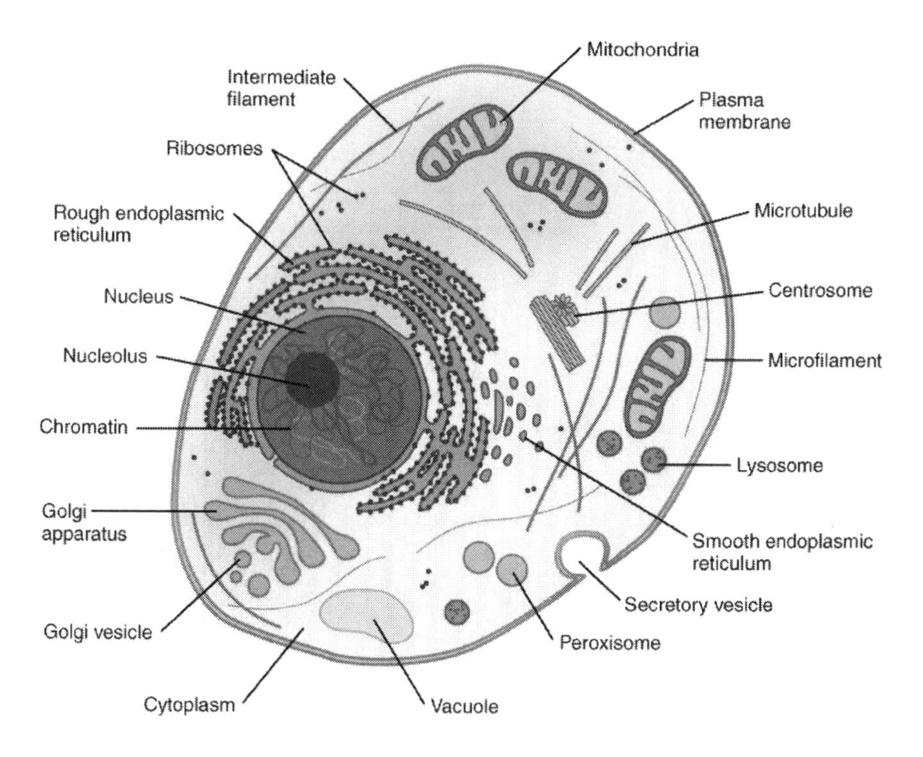

OTHER COMPONENTS AND STRUCTURES OF A CELL

Cell membrane	The 'gatekeeper' – the structure that encloses the cell.
Centrosomes	The structure that contains centrioles.
Chromatin	The material of which chromosomes of eukaryotes are made of. It consists of DNA, RNA, and protein.
Cytoplasm	The protoplasm that surrounds the nucleus.
Cytoskeleton	Cytoskeletal elements form a network of protein filaments

	and tubules in the cytoplasm.
Inclusions	These are nonfunctioning units which are sometimes found in the cytoplasm. Inclusions are usually temporary.
Microvilli	Microvilli increase the surface size of the cell, thereby increasing its absorptive capacity.
Nucleolus	The site of ribosomal RNA.
Ribonucleic Acid	Transfers genetic info to ribosomes.

Test Your Knowledge!

In the next section we will cover cell division and reproduction. To fully come to grips with that topic though, it is important that you remember the structures contained in a cell. **Can you name the components of the cell?**

14. ________________________________

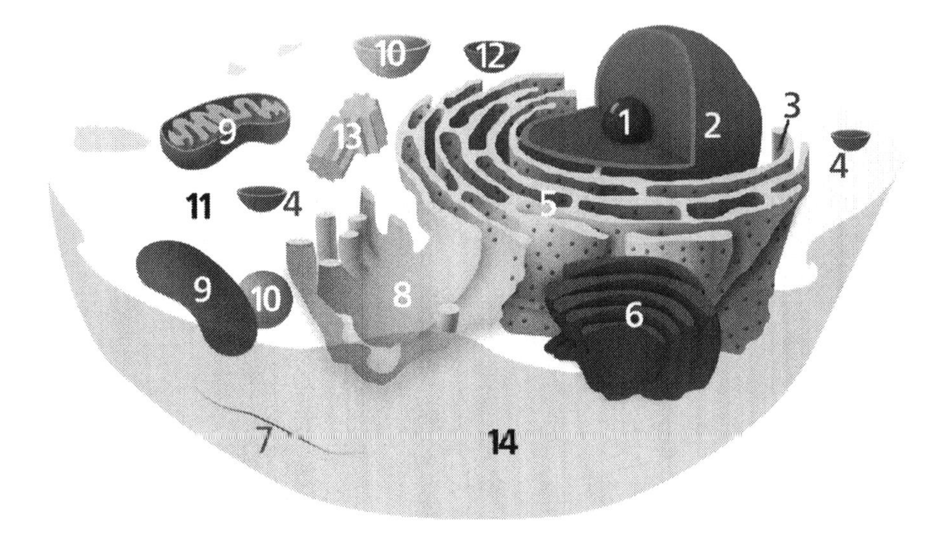

Answer:

1. Nucleolus
2. Nucleus
3. Ribosome (little dots)
4. Vesicle
5. Rough endoplasmic reticulum
6. Golgi apparatus
7. Cytoskeleton
8. Smooth endoplasmic reticulum
9. Mitochondrion
10. Vacuole
11. Cytosol (fluid that contains organelles)
12. Lysosome
13. Centrosome
14. Cell membrane

4. DNA, RNA, and Cell Reproduction

Protein synthesis is one of the most vital biological processes in which cells generate new proteins. *DNA (deoxyribonucleic acid)* carries genetic information, as well as provides the essential blueprint for *protein synthesis*. Protein synthesis is necessary for the repair of damaged tissues and for growth of new tissues. RNA (ribonucleic acid) is what transfers genetic information to the ribosomes, the site where protein synthesis occurs.

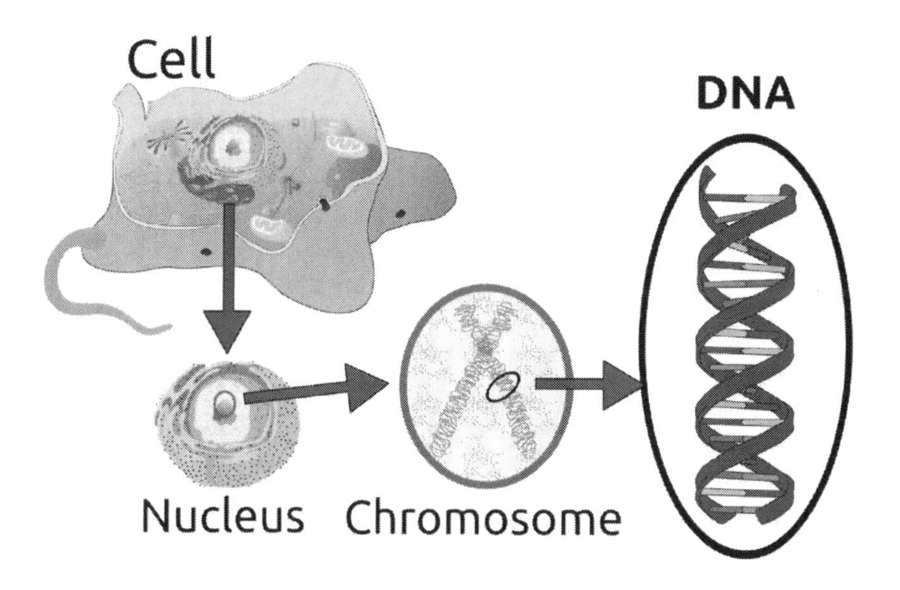

DNA is a nucleic acid, the basic structural units of which are nucleotides. Linked together, nucleotides form the building blocks of DNA or RNA. Each nucleotide is composed a nitrogen-containing nucleobase, a phosphate group, and a sugar called deoxyribose.

The four nucleobases that make up DNA are adenine (A), guanine (G), thymine (T), and cytosine (C). The nucleobases are classified into two types:

- Adenine and guanine are double-ring compounds classified as **'purines'**.
- Thymine and cytosine are single-ring compounds classified as **'pyrimidines'**.

Complementary Base Pairing

Nucleobases that link together are called **'complementary'**. Held together by a weak chemical attraction between the nitrogen bases, the two nucleotides bind together across the DNA double helix to form a **'base pair'**. Because of the chemical shape of the nucleobases however, adenine bonds only with thymine and guanine bonds only with cytosine.

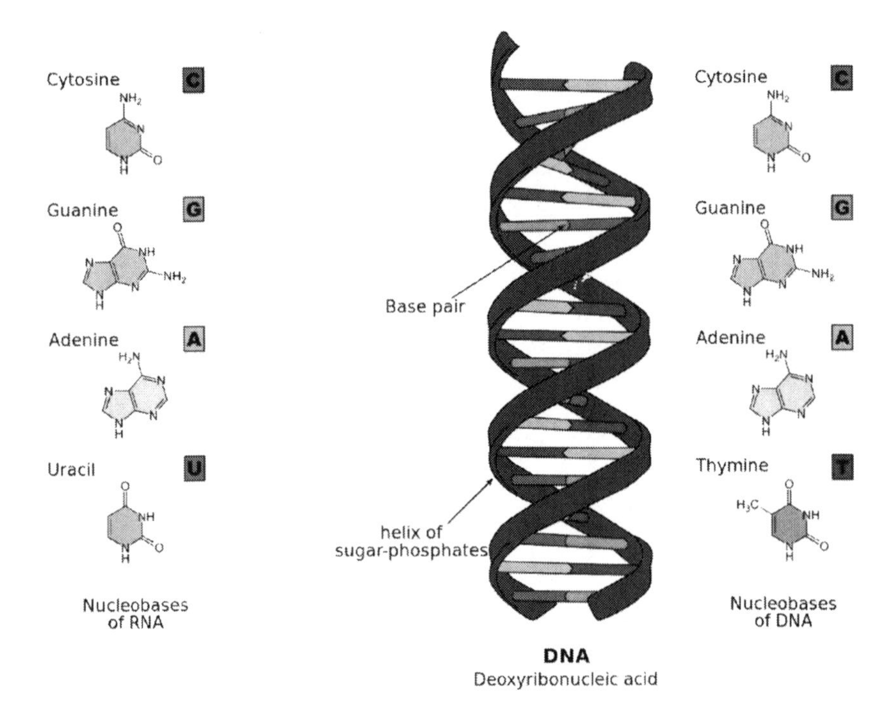

The nucleotides that make up RNA are slightly different to those found in DNA. Different types of RNA are involved in the transportation of genetic information to the ribosomes (the site of protein synthesis). These are: ribosomal RNA, messenger RNA and transfer RNA.

- **Ribosomal RNA (rRNA):** Ribosomal RNA are used to make ribosomes in the endoplasmic reticulum of the cytoplasm. Ribosomes are the site of protein synthesis.
- **Messenger RNA (mRNA):** Messenger RNA mediates the transfer of genetic information from the DNA to the ribosome, specifying the arrangement of amino acids to make proteins.
- **Transfer RNA (tRNA):** Transfer RNA helps decode the mRNA sequence into a protein. In other words, it transfers the genetic code from the mRNA for the production of a specific amino acid.

Cell Reproduction

Different cells have different lifespans and will all (excluding germ cells) experience aging (*senescence*) and death. The human body contains specialized cells that cannot divide and are therefore irreplaceable. Examples of these are muscle cells and nerve cells. In contrast to this, the body also has cells that reproduce at a fast rate to replace those cells with a short life span. These include the cells found in the outer layer of the skin. Lastly, there are cells that divide slowly under normal circumstances, but also rapidly if required. Such cells are found in connective tissue (repair after injury). In general, the simpler the cell, the greater its ability to regenerate. The more specialized a cell, the shorter its lifespan and the weaker its power to regenerate.

DNA Replication

Cell division is the process by which cells reproduce. Cell division can be achieved through the process of *mitosis* or *meiosis*. Before cell division can occur however, the chromosomes are replicated in a process called *DNA replication*. DNA replication occurs during interphase, prior to the beginning of mitosis or meiosis.

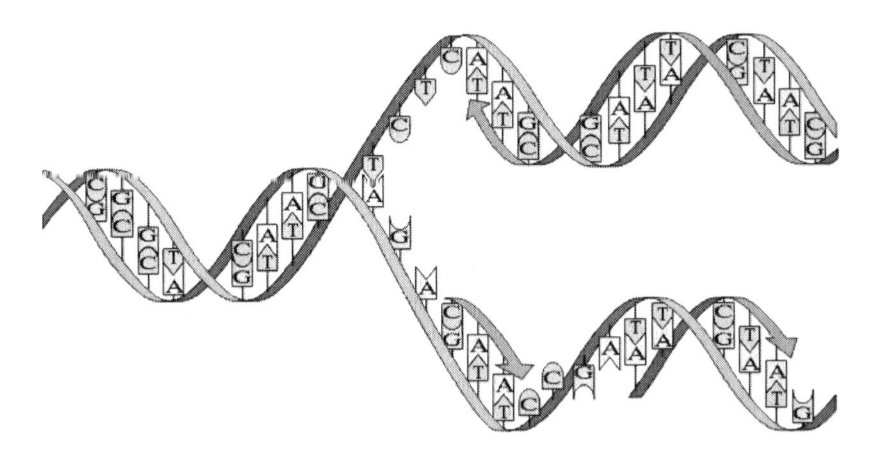

To replicate, the DNA double helix separates (*'unzips'*) into two separate DNA chains.

Each separate chain then acts as a 'template' for the construction of a new chain. New nucleotides are then added to the new strand, forming an identical double helix due to the fact that adenine bonds only with thymine and guanine bonds only with cytosine. Because of complementary base pairing, the two new double helixes (each of which contains one original strand and one newly formed strand) are therefore an exact duplicate of the original DNA double helix.

Mitosis

In the human body, mitosis is the preferred method of replication by all cells except for gametes. Mitosis is the division of a cell nucleus (***karyokinesis***), followed by a division of the cell body (***cytokinesis***), bringing about the separation into two *daughter cells.* In other words, the nuclear content of all cells of the body (except for gametes) reproduces and divides through the process of mitosis – the result of which is the formation of two new daughter cells. Each new daughter cell contains the *diploid* (46) number of chromosomes (i.e. 23 pairs of chromosomes).

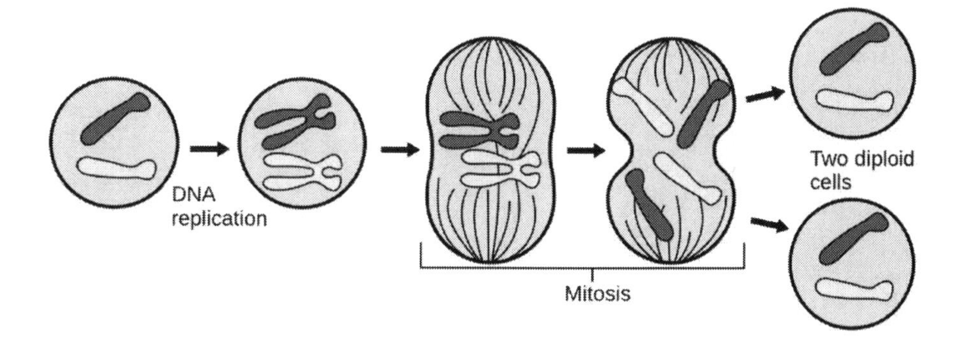

A cell enters mitosis in response to signals received via the mRNA from the nucleus. Mitosis comprises five stages, one of which is inactive (interphase) and four active phases:

1. **Interphase**: Interphase starts when the cell membrane fully encloses the new cell. During interphase, chromosomes replicate. Both the nuclear membrane and the nucleus are well defined.
2. **Prophase**: In prophase, the nucleus disappears, the nuclear membrane begins to disappear, and the chromosomes become more pronounced. Each duplicated chromosome is made up of two identical strands of DNA (chromatids – remember that once chromatids separate, each is considered to be a new chromosome). The chromatids remain attached by the **centromere**. The centrioles then migrate toward opposite poles of the cell, producing spindle fibers that extend to the midline (the equator) of the cell.

3. **Metaphase**: Metaphase occurs when the chromosomes line up at the equator of the cell between the spindles, on what is called the **metaphase plate**. At this stage, the centromeres divide.
4. **Anaphase**: In anaphase, the centromeres move apart (they are now referred to as chromosomes) and pull the sister chromatids separate and move to opposite poles.
5. **Telophase**: Telophase is the final stage of mitosis. During telophase, the spindle fibers disappear and a nuclear membrane forms around each nucleus. The cytoplasm compresses, dividing the cell in half. Each new cell contains a diploid (46) number of chromosomes identical to those in the original nucleus.

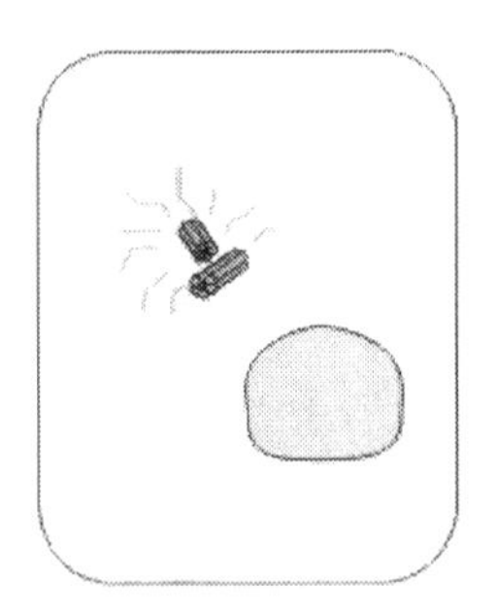

Interphase

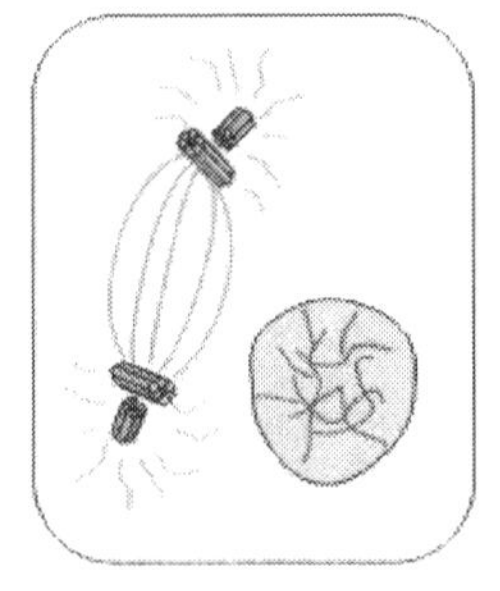

Early prophase

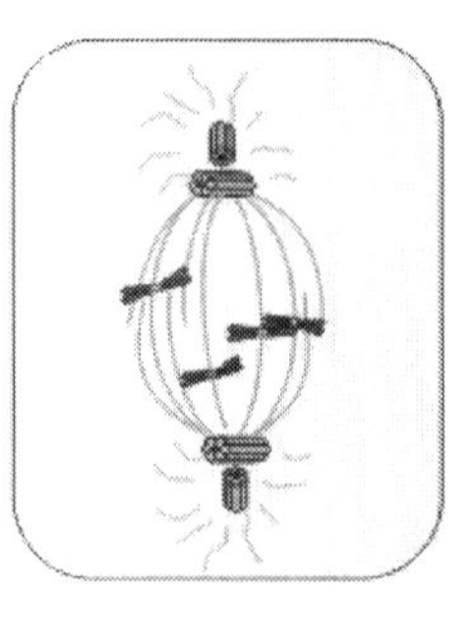

Late prophase

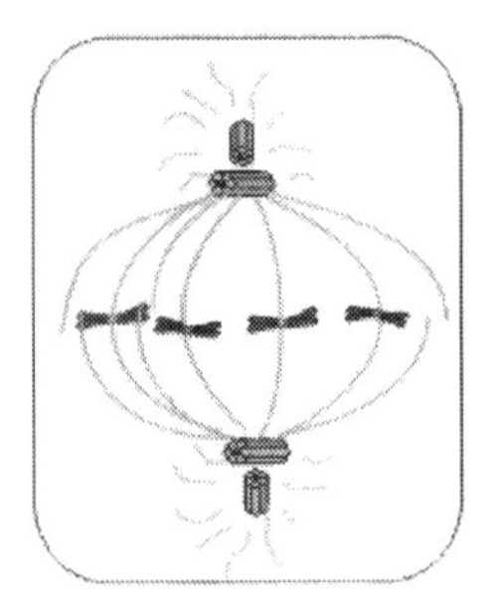

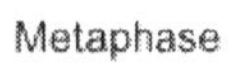

Metaphase

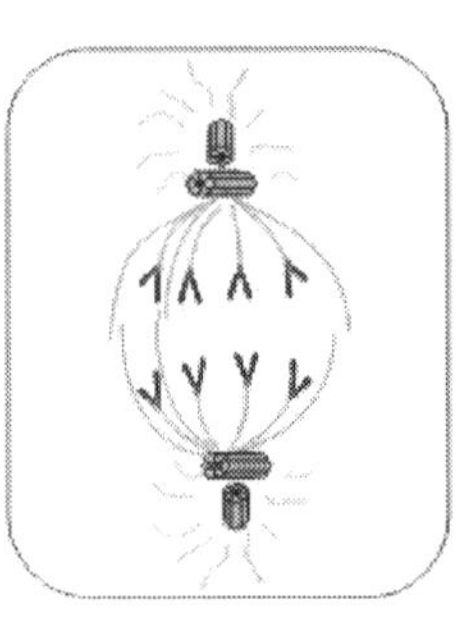

Anaphase

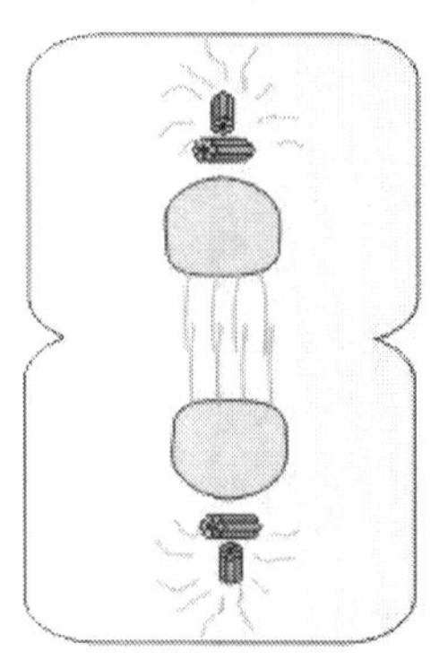

Telophase

Meiosis

A cell divides by mitosis for asexual reproduction and growth. The process of *meiosis* is reserved for gametes, also called sex cells. Gametes are of two kinds: ova and spermatozoa. Division by meiosis intermixes genetic material between homologous chromosomes - the result of which is the formation of four new daughter cells. Each new daughter cell contains the haploid (23) number of chromosomes.

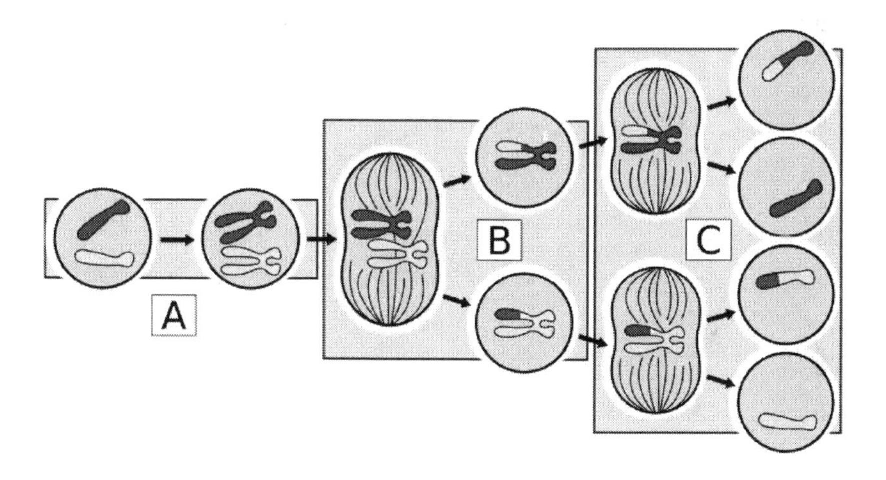

Like mitosis, division by way of meiosis begins with an interphase during which DNA replication occurs stage A in the diagram above). Unlike mitosis however, meiosis has two divisions, called **meiosis I** (stage B) and **meiosis II** (stage C). The two divisions are separated by a resting phase. The first division is characterized by six phases and the second division has four phases.

THE FIRST DIVISION: MEIOSIS I

1. **Interphase**

During interphase, the DNA replication occurs. The chromosomes are not yet distinct in appearance. Both the nuclear membrane and the nucleus are well defined.

2. **Prophase I**

The nucleus disappears and the nuclear membrane begins to disappear. Homologous chromosomes condense, move closer together and exchange genetic information. Genetic recombination may occur, which can result in a new set of genetic information. The centrioles migrate toward opposite poles of the cell, forming spindle fibers between them.

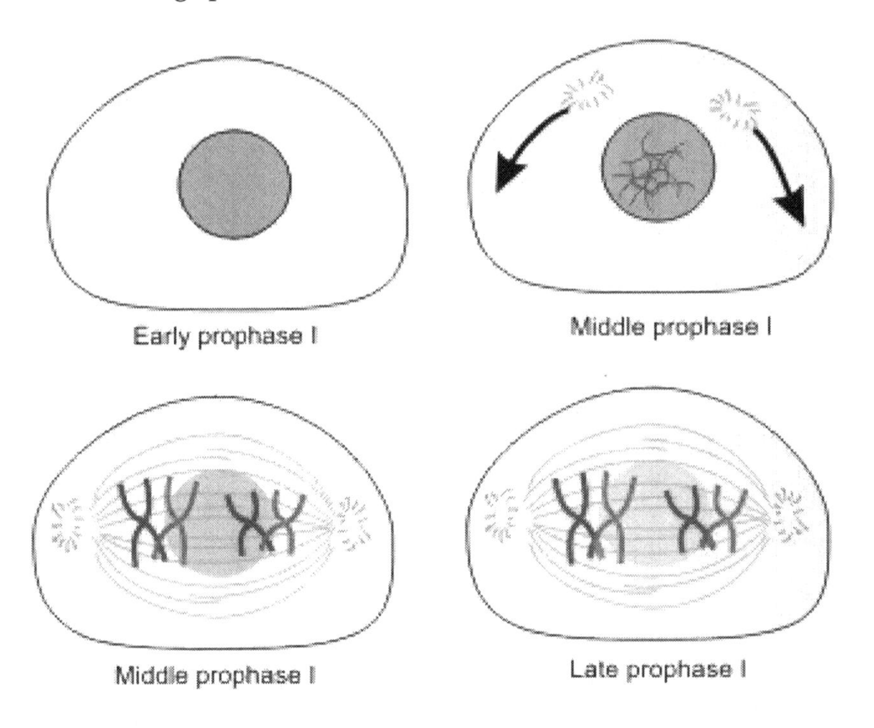

3. **Metaphase I**

Synaptic chromosome pairs align along the metaphase plate and the spindle apparatus attaches to the chromosomes.

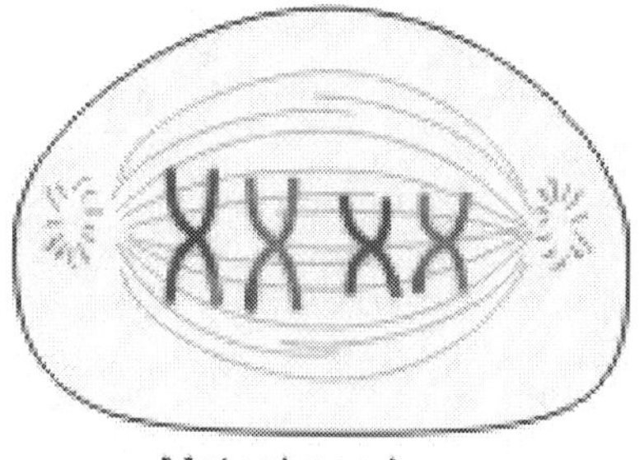

Metaphase I

4. **Anaphase I**

During anaphase I, the synaptic pairs of chromosomes separate. The spindle fibers pull the homologous double-stranded chromosomes to opposite poles of the cell. Centromeres do not divide.

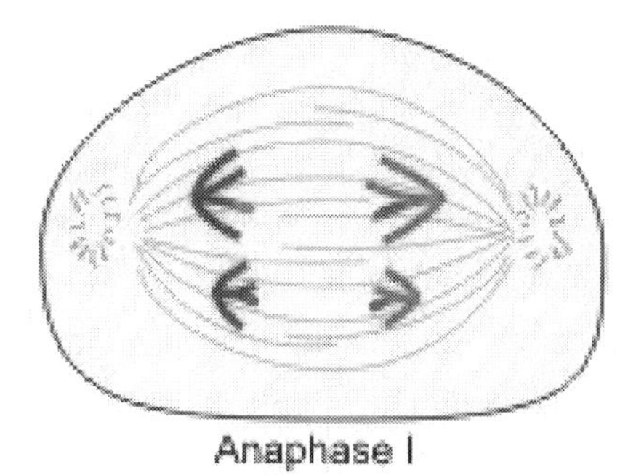

Anaphase I

5. **Telophase I**

Nuclear division begins – the nuclear membrane forms and the cytoplasm compresses, dividing the cell into halves. The chromosomes and spindle fibers disappear. Two new daughter cells are formed, each containing the haploids (23) number of chromosomes.

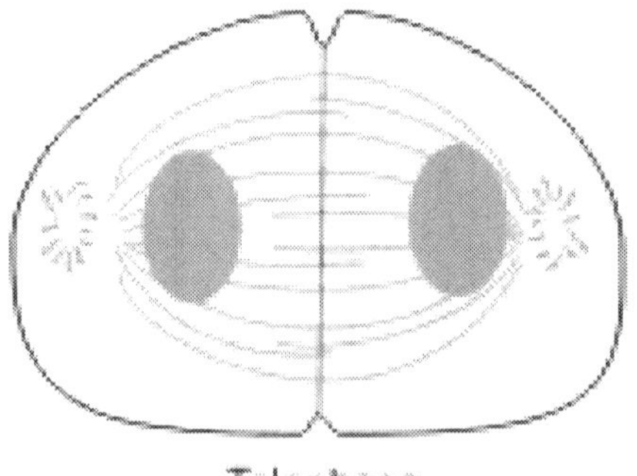

Telophase

6. **Interkinesis**

Interkinesis is the resting phase between meiosis I and meiosis II. During interkinesis, the nucleus and the nuclear membrane are well defined.

The Second Division: **Meiosis II**

1. **Prophase II**

During prophase II, the nuclear membrane disappears and the chromosomes condense. Spindle fibers begin to form between the centrioles, which begin to migrate toward opposite poles.

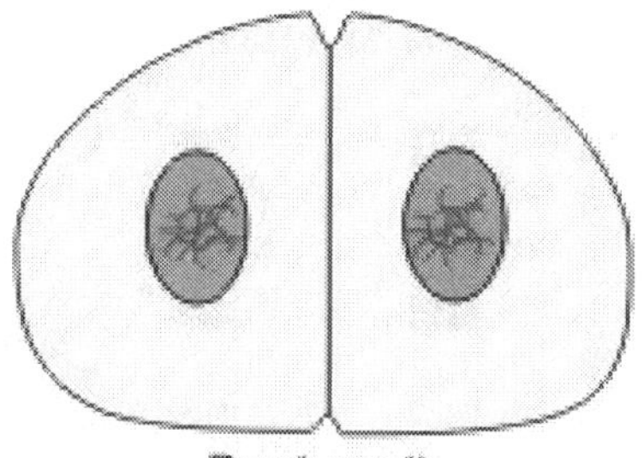

Prophase II

2. **Metaphase II**
3.

During metaphase II, the chromosomes line up along the metaphase plate and the spindle apparatus attaches to the chromosomes.

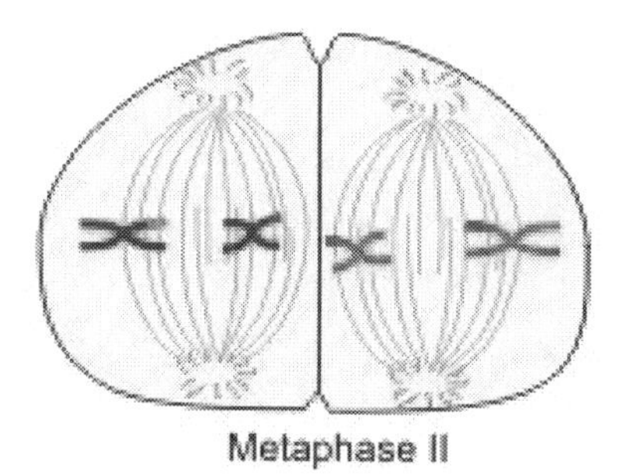

Metaphase II

4. **Anaphase II**

During anaphase II, sister chromatids separate into single-stranded chromosomes. These single-stranded chromosomes begin to migrate toward opposite poles.

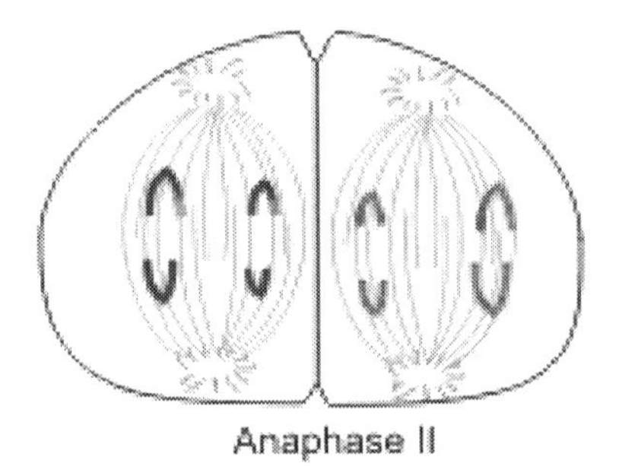

5. **Telophase II**

During telophase II, the nuclear membrane forms and the chromosomes and spindle fibers disappear. The cytoplasm compresses, dividing in halves. Cell division completes. Four new daughter cells are created, each containing the haploids (23) number of chromosomes.

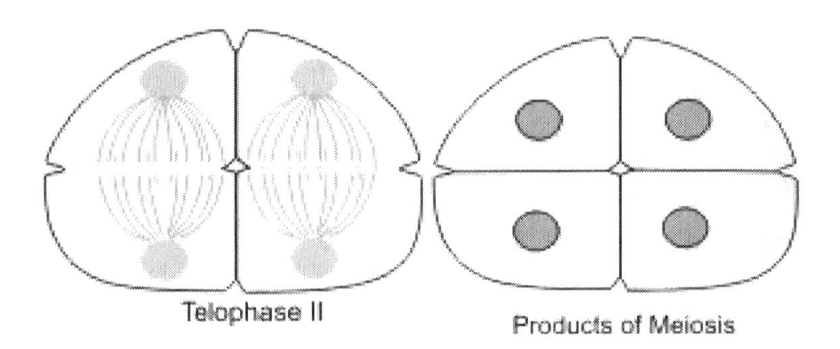

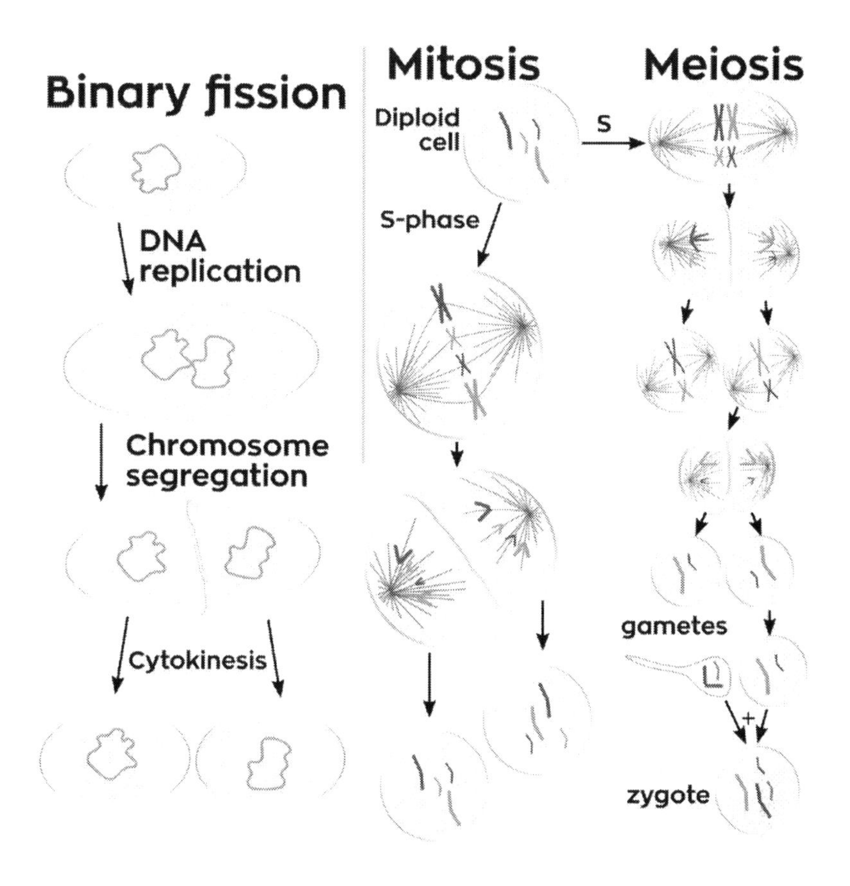
Binary fission
DNA replication
Chromosome segregation
Cytokinesis
Mitosis
Diploid cell
S-phase
Meiosis
S
gametes
+
zygote

5. Human Tissue

Tissues are groups of cells that are morphologically similar and that act together to perform specific functions in the body. The human body contains four basic types of tissue: epithelial tissue, *connective tissue, muscle tissue,* and *nerve tissue.*

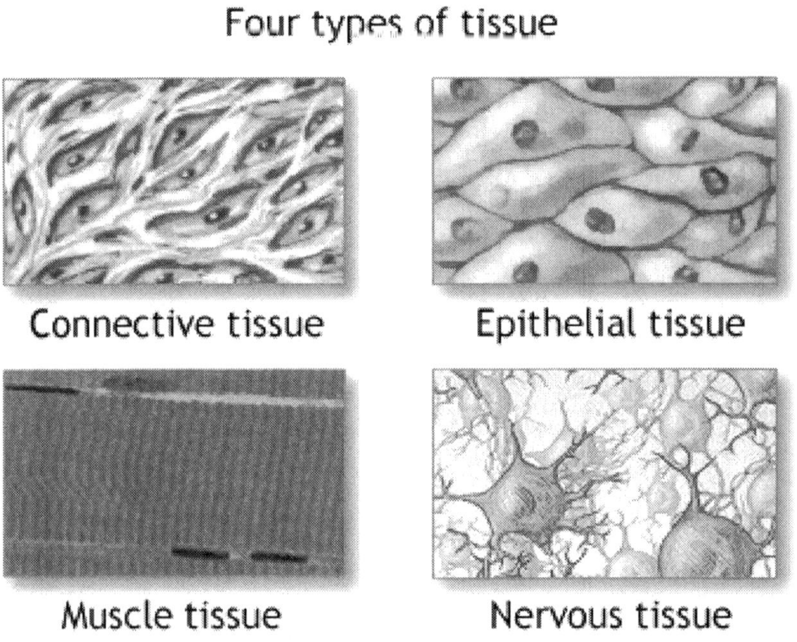

Four types of tissue

Connective tissue

Epithelial tissue

Muscle tissue

Nervous tissue

1. ***Epithelial tissue***

Epithelial tissue or epithelium is the thin tissue that covers the body's surface, i.e. the outer layer of the skin is made up of epithelial tissue. The epithelium also lines the inside of organs within the body and the body cavities. Epithelial tissue is classified in two ways: by the number of cell layers and the shape of the surface cells.

Number of Cell Layers

Classified by the number of cell layers, epithelium can be simple, stratified or pseudostratified (an epithelium that gives the superficial appearance of being stratified):

- Simple = One-layered
- Stratified = Multilayered
- Pseudostratified = One-layered but appearing to be multilayered

Cell shape

Classified by the shape of the surface cells, epithelium may be squamous, columnar, or cuboidal:

- Squamous = tissue contains flat surface cells
- Columnar = tissue contains tall surface cells that appear cylindrical in appearance
- Cuboidal = tissue contains cube-shaped surface cells

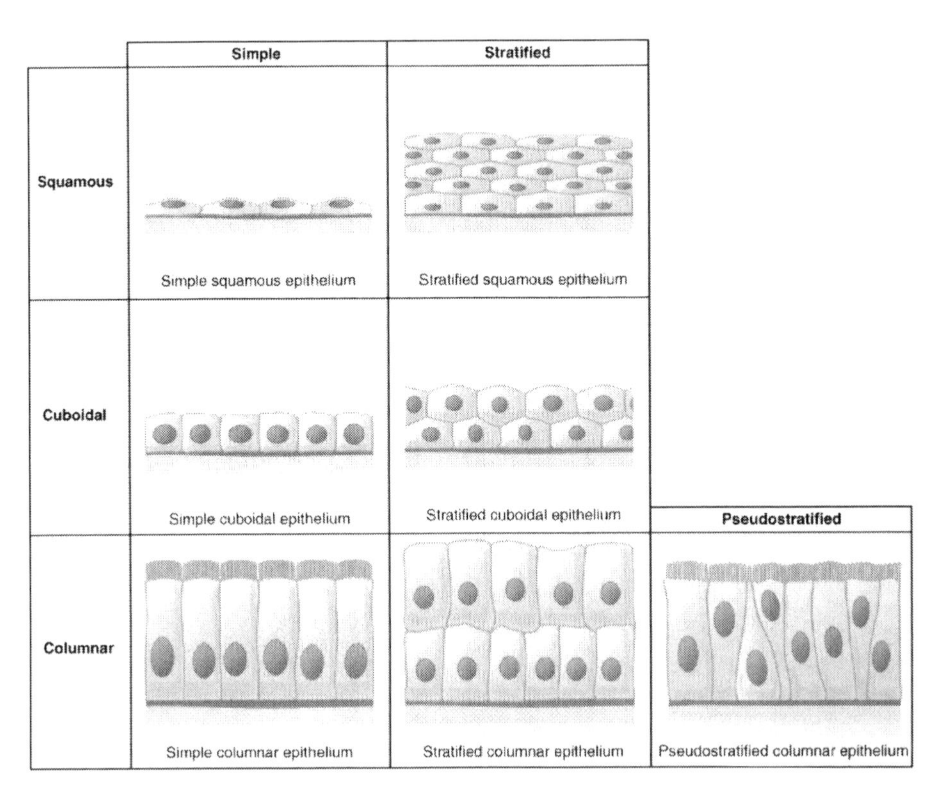

Borders of Columnar Epithelial Cells

Some columnar epithelial cells have vertical striations (a series of ridges), which form a ***striated border.*** Epithelial cells with a striated border are found in the lining of the intestines for example.

Some epithelial cells that have brush-like structures on their surface (microvilli) have a ***brush border.*** Epithelial cells with a brush border are found in the tubules of the kidneys for example.

Endothelium

Epithelial tissue (also called epithelium) that is made up of a single layer of squamous cells is called **endothelium.** Endothelium tissue lines the heart, blood vessels, and lymphatic vessels.

Cells	Location	Function
Simple squamous epithelium	Air sacs of lungs and the lining of the heart, blood vessels, and lymphatic vessels	Allows materials to pass through by diffusion and filtration, and secretes lubricating substance
Simple cuboidal epithelium	In ducts and secretory portions of small glands and in kidney tubules	Secretes and absorbs
Simple columnar epithelium	Ciliated tissues are in bronchi, uterine tubes, and uterus; smooth (nonciliated tissues) are in the digestive tract, bladder	Absorbs; it also secretes mucous and enzymes
Pseudostratified columnar epithelium	Ciliated tissue lines the trachea and much of the upper respiratory tract	Secretes mucus; ciliated tissue moves mucus
Stratified squamous epithelium	Lines the esophagus, mouth, and vagina	Protects against abrasion
Stratified cuboidal epithelium	Sweat glands, salivary glands, and the mammary glands	Protective tissue

Stratified columnar epithelium 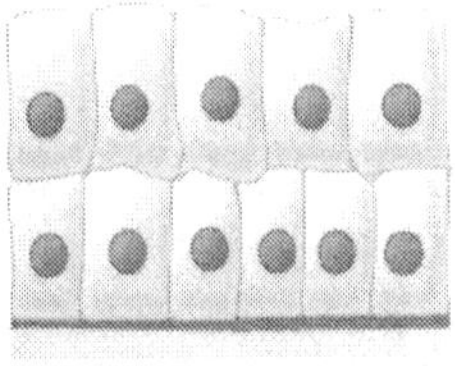	The male urethra and the ducts of some glands	Secretes and protects

2. <u>*Connective Tissue*</u>

Connective tissue connects, supports, and binds body structures. It also joins, nourishes, insulates, and protects organs. Connective tissue is classified as being ***loose*** or ***tense.***

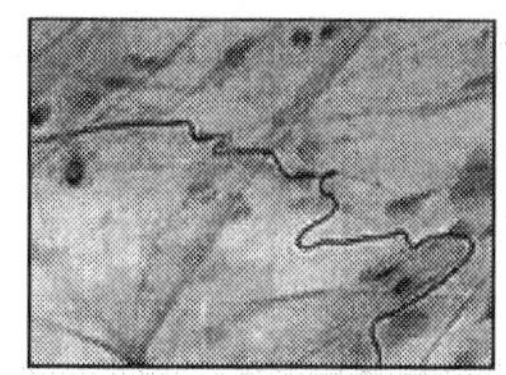

Areolar connective tissue

Adipose tissue

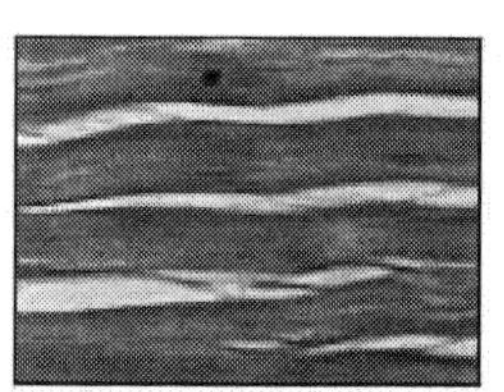

Fibrous connective tissue

Loose (areolar) connective tissue contains a lot of intercellular fluid and has large spaces that separate cells and fibers.

Adipose tissue (*fat*) is a special type of loose connective tissue composed of mostly adipocytes. Adipocytes are cells that specialize in the storage of lipids (fats).

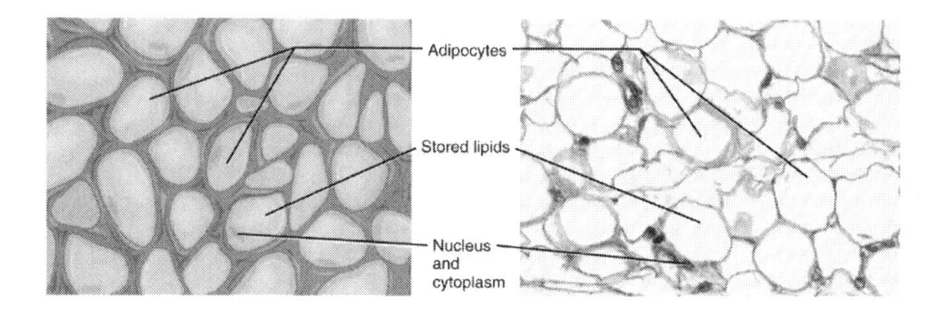

Dense (fibrous) connective tissue has a great fiber concentration and provides structural support. Dense connective tissue is further classified into dense regular and irregular dense.

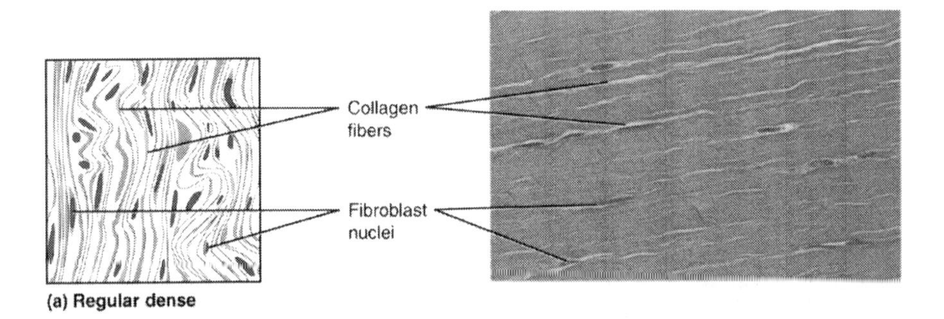

(a) Regular dense

Regular dense connective tissue is made up of tightly packed fibers that form a consistent pattern (the pattern is 'regular'). Tendons, ligaments and aponeuroses consist of regular dense connective tissue.

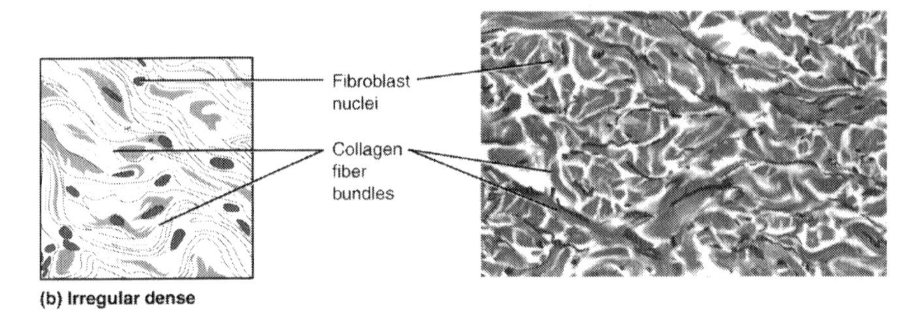

(b) Irregular dense

Irregular dense connective tissue is made up tightly packed fibers that create an inconsistent pattern (the pattern is 'irregular'). Irregular dense connective tissue can be found in the fasciae, the submucosa of the GI tract, fibrous capsules, and the dermis.

3. ***Muscle tissue***

Muscle tissues are made up of muscle cells that contract and relax to produce movement. Muscle cells are elongated in shape in order to facilitate contractibility and benefit from generous blood supply.

There are three types of muscle tissue: ***skeletal (striated) muscle tissue, smooth-muscle tissue, cardiac muscle tissue.***

1. **Skeletal (Striated) Muscle Tissue**

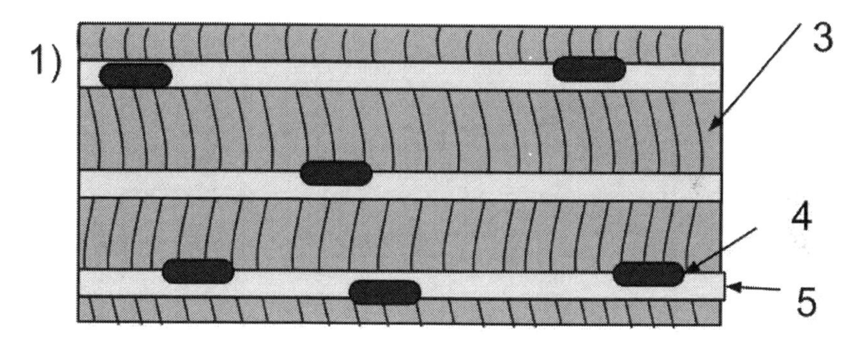

Skeletal muscle tissue contracts voluntarily. This tissue type is found in the muscles that are attached to the skeleton. Skeletal muscle tissue is made up of long striated tubular cells (see 3 in the above diagram) and multiple nuclei (4). The nuclei are embedded in the plasma membrane (5).

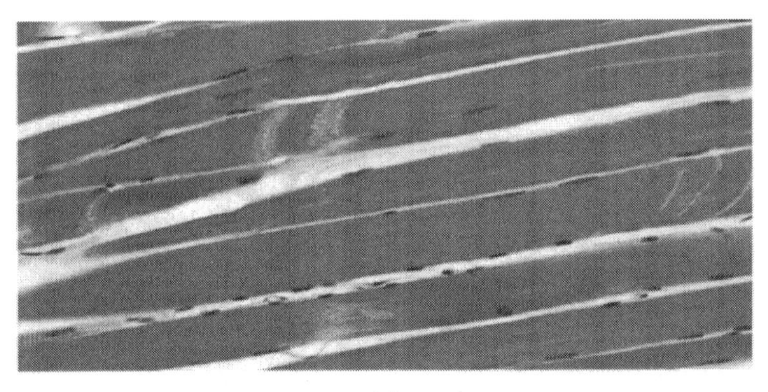

(a)

2. **Smooth Muscle Tissue**

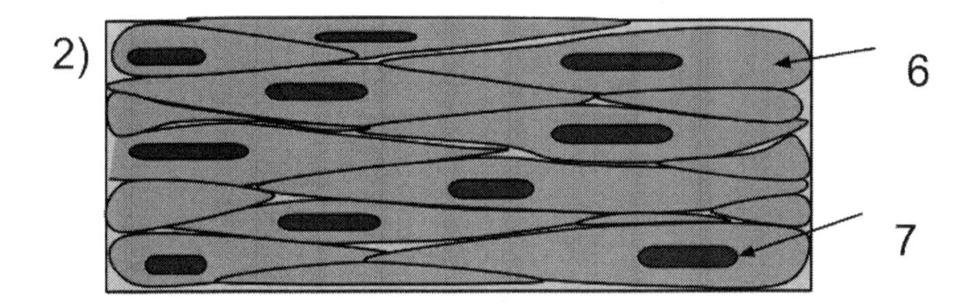

Smooth muscle tissue is primarily found in the digestive tract and in the walls of blood vessels, but it also lines the walls of any internal organs and other bodily structures. This type of muscle tissue consists of long and spindle-shaped cells (6) each of which contains its own nucleus (7). Smooth muscle tissue is non-striated and involuntary with its contractions being stimulated from the autonomous nervous system.

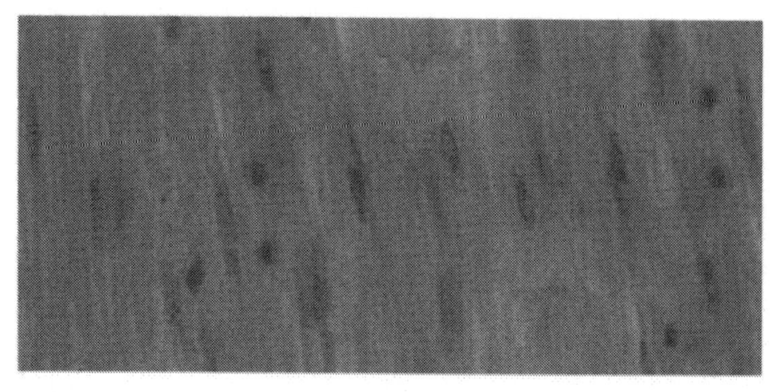

(b)

3. **Cardiac Muscle Tissue**

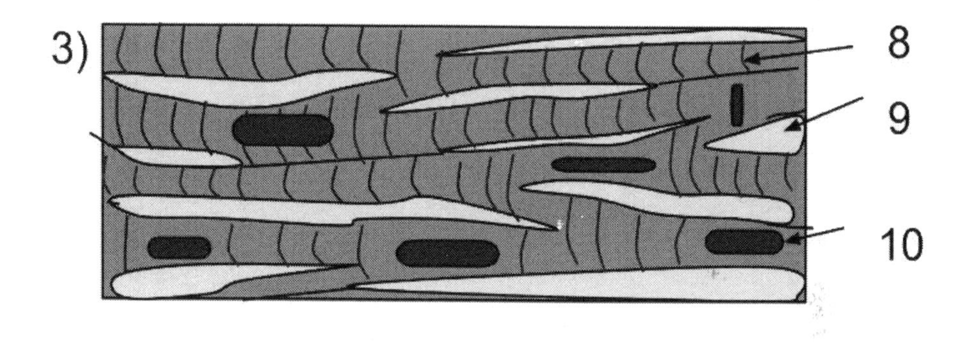

Cardiac muscle tissue is made up of cells that have striations (8), but differs from skeletal (striated) muscle tissue in two ways: firstly, its contractions are involuntary and secondly, its fibers are separate cellular units. Cardiac muscle cells each contain a single nucleus (10) and branch off from each other. This branching off creates junctions (9) between adjacent cells. Cardiac muscle tissue is only found in the heart and its primary function is to pump blood.

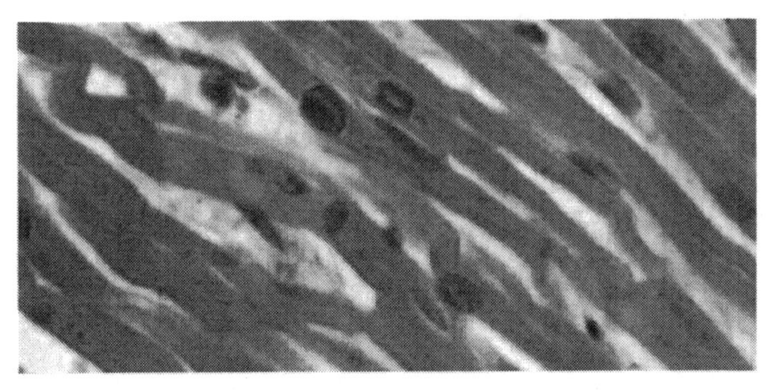

(c)

4. *Nervous tissue*

The primary function of nervous tissue is communication, to transmit information in the form of nerve impulses through the body. Nervous tissue is found in the brain, nerves, and sense organs and has two main properties:

- **Irritability**: the ability to respond to a stimulus, to react to an array of different physical and chemical agents.
- **Conductivity**: the ability to transmit the 'response' (the resulting action) from one point to another.

Nervous tissue is composed of **neurons** (nerve cells) and **neuroglia**. Neurons are highly specialized cells that generate and conduct nerve impulses. The anatomy of neurons and how neurons operate to receive and transmit impulses is covered in ***section 6*** (the neurosensory system) in greater depth. Neuroglia is the support structure of nervous tissue which is only found in the central nervous system. Neuroglia nourishes, insulates, and protects neurons. It also assists in the propagation of nerve impulses.

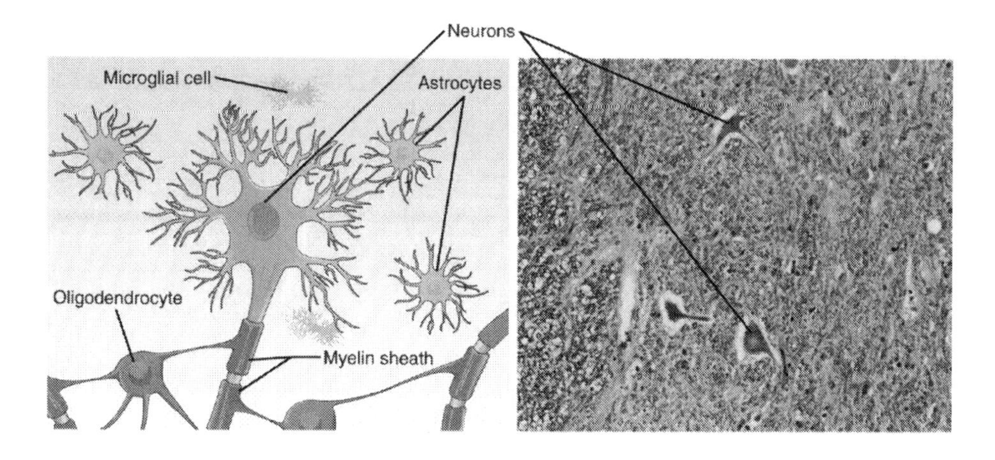
Neurons
Microglial cell
Astrocytes
Oligodendrocyte
Myelin sheath

Section 4: Structural Organization and Essential Medical Terminology

Structural Organization of the Human Body

The human body contains six levels of structural organization: ***the chemical level, the cellular level, the tissue level, the organ level, the organ system level, and the organism level.***

1. **Chemical Level**

The first and simplest level in the body's structural hierarchy is the chemical level. The chemical level comprises the smallest building blocks which combine to form molecules. These molecules in turn combine to form the organelles found in cells.

The body is composed of cells that consist of thousands of different chemicals elements, but four of these (***oxygen, carbon, hydrogen, and nitrogen***) make up 96.2% of the human body.

Element	Symbol	Percentage in Body
Oxygen	O	65.0
Carbon	C	18.5
Hydrogen	H	9.5
Nitrogen	N	3.2
Calcium	Ca	1.5
Phosphorus	P	1.0
Potassium	K	0.4
Sulfur	S	0.3
Sodium	Na	0.2
Chlorine	Cl	0.2
Magnesium	Mg	0.1
Trace elements include boron (B), chromium (Cr), cobalt (Co), copper (Cu), fluorine (F), iodine (I), iron (Fe), manganese (Mn), molybdenum (Mo), selenium (Se), silicon (Si), tin (Sn), vanadium (V), and zinc (Zn).		less than 1.0

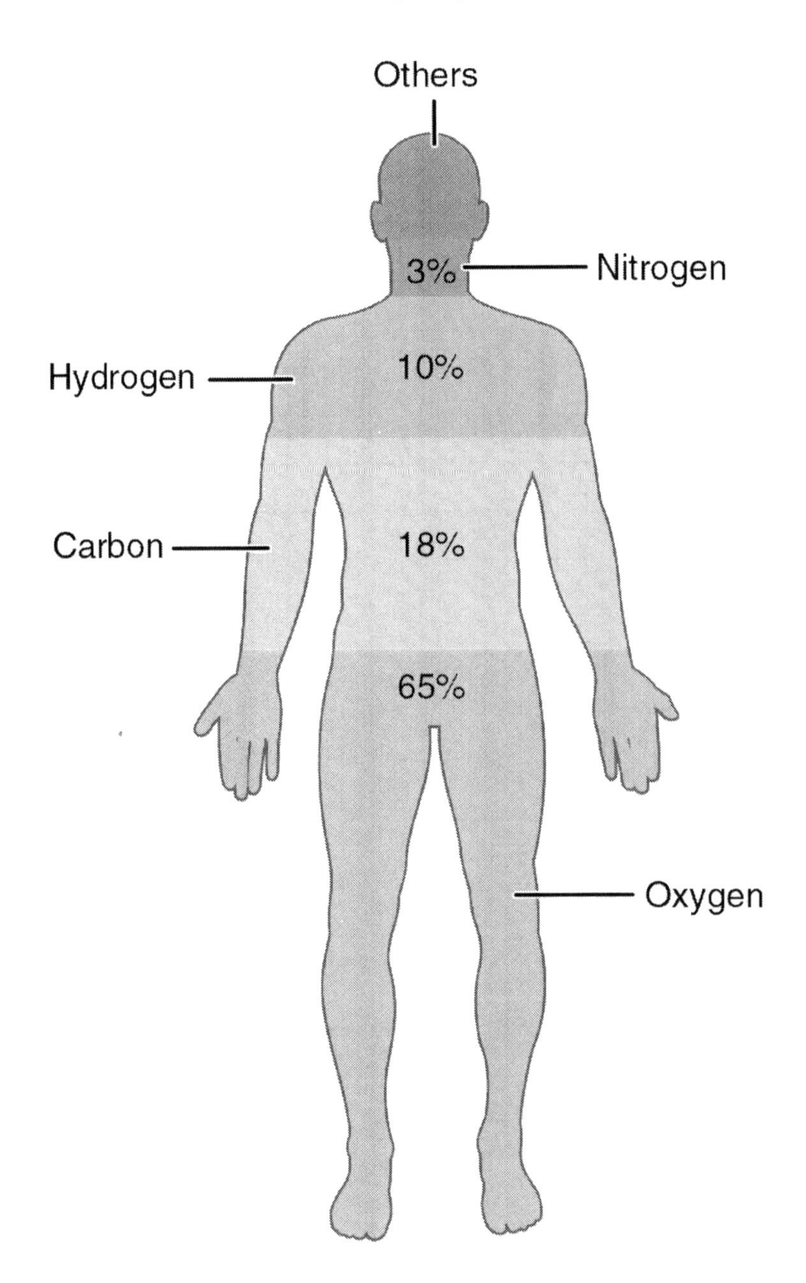

2. **Cellular Level**

The cellular level comprises the smallest unit of living organisms: the cell, each of which is carries out a set of specific tasks within the human body.

3. **Tissue Level**

Cells that share a common function group together to form tissues. The four main types of tissue in humans include ***epithelium, connective, muscle, and nervous tissue.***

4. **Organ Level**

An organ is a structure found in the body which performs a specific and typically more complex function within the body and which is composed of at least two different types of tissue.

5. **Organ System Level**

An organ system consists of a group of organs that work together to perform one or more functions. *(This book is structured according to the organ system level).*

6. **Organism Level**

The organism level is the highest level of organization - an individual form of life, such as the human being.

Anatomical Terms

Throughout this book, reference is made to anatomical

terminology to describe the anatomical location and action of different body structures.

- **Acromial:** Point of the shoulder
- **Antebrachial:** Forearm
- **Axillary:** Armpit
- **Brachial**: Arm
- **Buccal:** Cheek
- **Carpal:** Wrist
- **Celiac:** Abdomen
- **Cephalic:** Head
- **Cervical**: Neck
- **Costal:** Ribs
- **Coxal:** Hip
- **Crural:** Leg
- **Cubital:** Elbow
- **Digital:** Finger
- **Dorsum:** Back
- **Femoral:** Thigh
- **Frontal:** Forehead
- **Genital:** Reproductive organs
- **Gluteal:** Buttocks
- **Mental:** Chin
- **Nasal:** Nose
- **Oral:** Mouth
- **Otic:** Ear
- **Palmar:** Palm of the hand
- **Patellar**: Kneecap or kneepan
- **Pedal**: Foot
- **Pelvic:** Pelvis
- **Plantar:** Sole of the foot
- **Tarsal:** Tarsus of the foot, instep of the foot
- **Umbilical:** Navel
- **Vertebral:** Spinal column

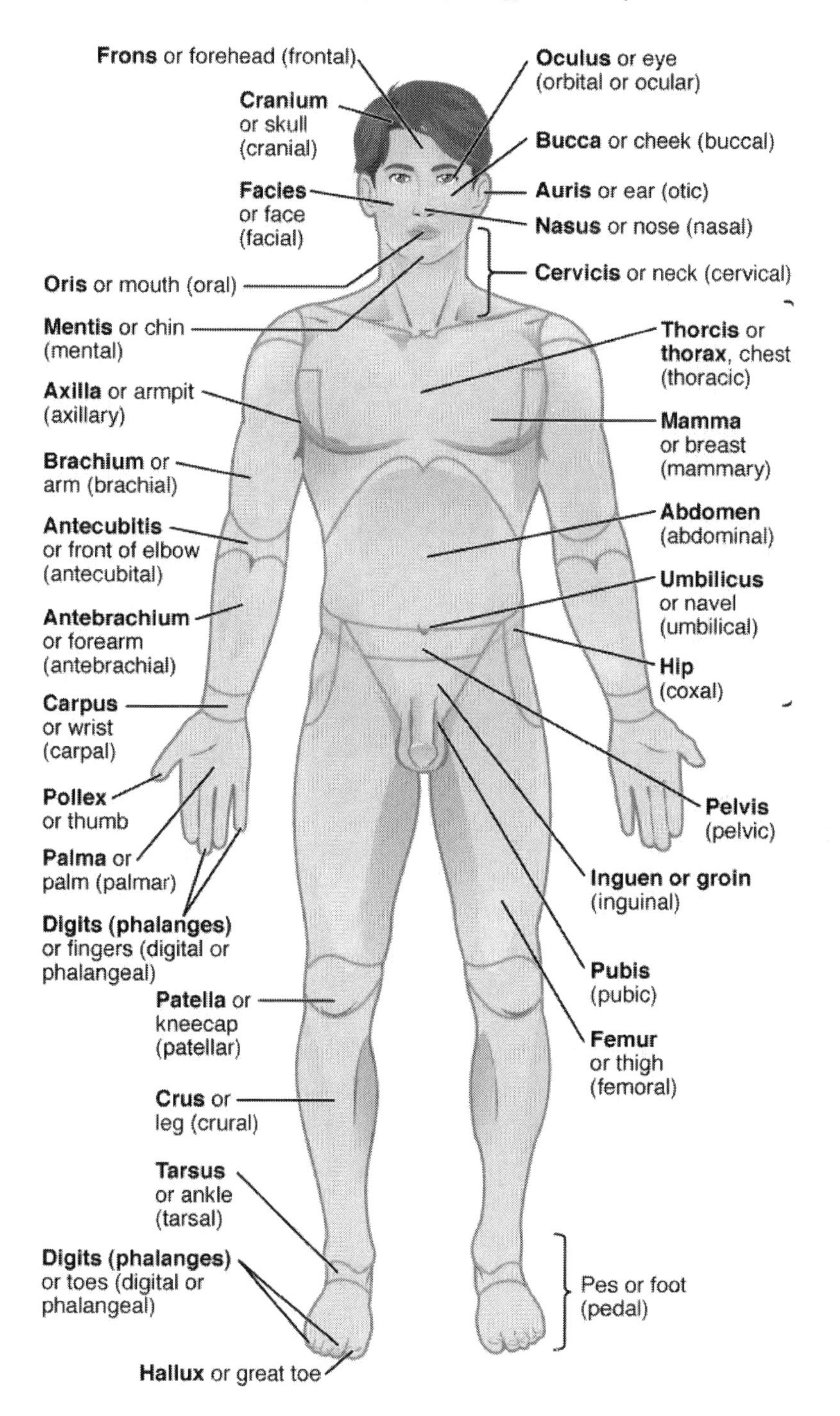
Frons or forehead (frontal)
Cranium or skull (cranial)
Facies or face (facial)
Oris or mouth (oral)
Mentis or chin (mental)
Axilla or armpit (axillary)
Brachium or arm (brachial)
Antecubitis or front of elbow (antecubital)
Antebrachium or forearm (antebrachial)
Carpus or wrist (carpal)
Pollex or thumb
Palma or palm (palmar)
Digits (phalanges) or fingers (digital or phalangeal)
Patella or kneecap (patellar)
Crus or leg (crural)
Tarsus or ankle (tarsal)
Digits (phalanges) or toes (digital or phalangeal)
Hallux or great toe
Oculus or eye (orbital or ocular)
Bucca or cheek (buccal)
Auris or ear (otic)
Nasus or nose (nasal)
Cervicis or neck (cervical)
Thorcis or thorax, chest (thoracic)
Mamma or breast (mammary)
Abdomen (abdominal)
Umbilicus or navel (umbilical)
Hip (coxal)
Pelvis (pelvic)
Inguen or groin (inguinal)
Pubis (pubic)
Femur or thigh (femoral)
Pes or foot (pedal)

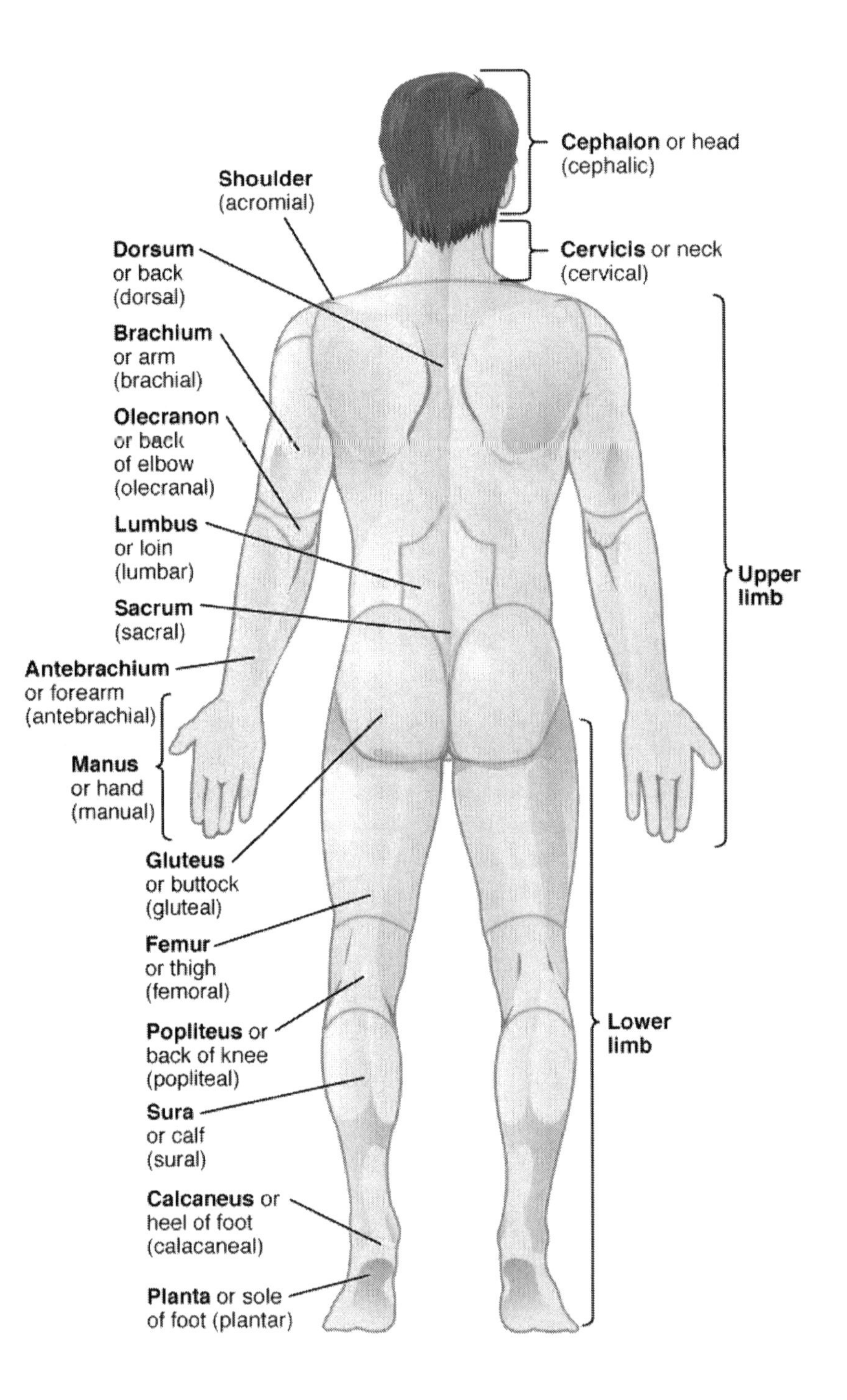
Cephalon or head
(cephalic)
Shoulder
(acromial)
Dorsum
or back
(dorsal)
Cervicis or neck
(cervical)
Brachium
or arm
(brachial)
Olecranon
or back
of elbow
(olecranal)
Lumbus
or loin
(lumbar)
Upper
limb
Sacrum
(sacral)
Antebrachium
or forearm
(antebrachial)
Manus
or hand
(manual)
Gluteus
or buttock
(gluteal)
Femur
or thigh
(femoral)
Popliteus or
back of knee
(popliteal)
Lower
limb
Sura
or calf
(sural)
Calcaneus or
heel of foot
(calacaneal)
Planta or sole
of foot (plantar)

Essential Anatomical Terms:

- **Acromial:** Point of the shoulder
- **Antebrachial:** Forearm
- **Axillary:** Armpit
- **Brachial:** Arm
- **Buccal:** Cheek
- **Carpal:** Wrist
- **Celiac:** Abdomen
- **Cephalic:** Head
- **Cervical:** Neck
- **Costal:** Ribs
- **Coxal:** Hip
- **Crural:** Leg
- **Cubital:** Elbow
- **Digital:** Finger
- **Dorsum:** Back
- **Femoral:** Thigh
- **Frontal:** Forehead
- **Genital:** Reproductive organs
- **Gluteal:** Buttocks
- **Mental:** Chin
- **Nasal:** Nose
- **Oral:** Mouth
- **Otic:** Ear
- **Palmar:** Palm of the hand
- **Patellar:** Kneecap or kneepan
- **Pedal:** Foot
- **Pelvic:** Pelvis
- **Plantar:** Sole of the foot
- **Tarsal:** Tarsus of the foot, instep of the foot
- **Umbilical:** Navel
- **Vertebral:** Spinal column

Anatomical Root Words

Root Word = Meaning

Skeletal System

Os-, Oste- = Bone

Arth- = Joint

Muscular System

Myo- = Muscle

Sarco- = Striated muscle

Integument

Derm- = Skin

Nervous System

Neur- = Nerve

Endocrine System

Aden- = Gland

Estr- = Steroid

Circulatory System

Card- = Heart muscle

Angi- = Vessel

Hema- = Blood

Arter- = Artery

Ven- = Venous

Erythro- = Red

Respiratory System

Pulmon- = Lung

Bronch- = Windpipe

Digestive System

Gastr- = Stomach

Enter- = Intestine

Dent- = Teeth

Hepat- = Liver

Urinary System

Ren-, Neph- = Kidney

Ur- = Urinary

Immune System

Lymph- = Lymph

Leuk- = White

Reproductive System

Vagin- = Vagina

Uter- = Uterine

Anatomical Root Words

Root Word = Meaning

Skeletal System

Os-, Oste- = Bone

Arth- = Joint

Muscular System

Myo- = Muscle

Sarco- = Striated muscle

Integument

Derm- = Skin

Nervous System

Neur- = Nerve

Endocrine System

Aden- = Gland

Estr- = Steroid

Circulatory System

Card- = Heart muscle

Angi- = Vessel

Hema- = Blood

Arter- = Artery

Ven- = Venous

Erythro- = Red

Respiratory System

Pulmon- = Lung

Bronch- = Windpipe

Digestive System

Gastr- = Stomach

Enter- = Intestine

Dent- = Teeth

Hepat- = Liver

Urinary System

Ren-, Neph- = Kidney

Ur- = Urinary

Immune System

Lymph- = Lymph

Leuk- = White

Reproductive System

Vagin- = Vagina

Uter- = Uterine

An understanding of the following directional prefixes is important to navigate human anatomy:

Prefix = Meaning

Ab- = Away from

Ad- = Toward

Circum- = Around

Contra- = Opposition, against

De- = Down, away from

Ecto-, Exo- = Outside

Endo- = Within

Epi- = Upon, over

Extra- = Outside

Infra- (sub) = Below, under

Intra- = Inside

Ipsi- (iso) = Same (equal)

Ir- = Into, toward

Meso- = Middle

Meta- (supra) = Beyond, over, after

Para- = Near, beside

Peri- = Around, surrounding

Retro- = Behind, backward

Sub- = Under, rear,

Trans- = Across, through

Section 5: The Musculoskeletal System

The musculoskeletal system consists of the structures that determine the shape of the human body and that allow it to move. It consists of the bones that make up the skeleton, the muscles that support the body, and a number of other structures, such as tendons, ligaments, cartilage, and the joints, which help bind and support structures of the body. The primary role of the musculoskeletal system is to support the body, protect internal structures and vital organ, and facilitate movement.

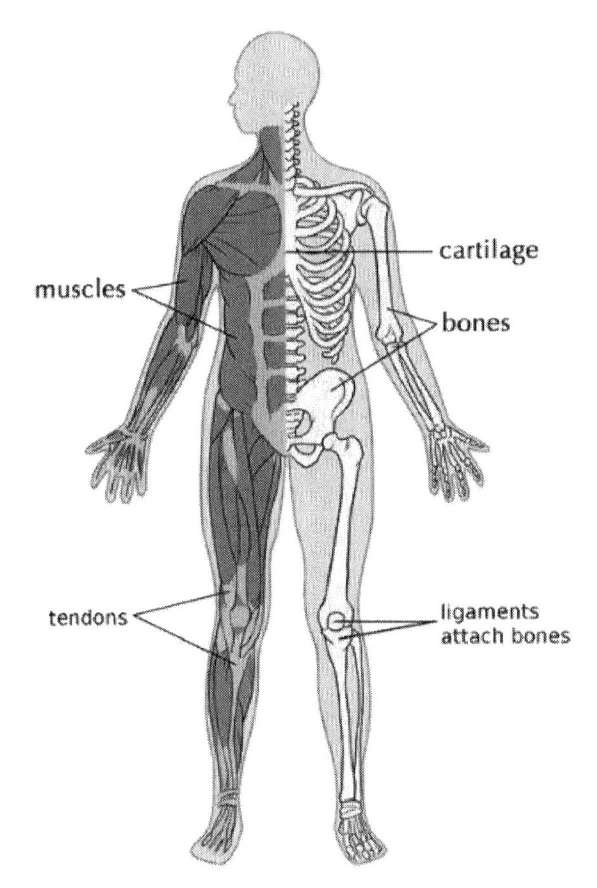

This section will begin by looking at the major muscles of the body, muscle structure, attachment, and structure. We will then cover the major bones of the body, bone structure, growth, and remodeling. Lastly, we will cover other important musculoskeletal structures that support and bind tissues and organs together, such as the joints and cartilage.

The Muscular System

The human body comprises around 600 skeletal muscles that are attached to the bones by tendons, which are bundles of collagen fibers. These muscles hold the body together, maintain posture, generate body heat, and protect the organs.

Just as there are three types of muscle tissue (see *section 3*), there are three types of muscle in the human body: ***the skeletal (voluntary and reflex) muscle, the smooth (involuntary) muscle, and the cardiac (heart) muscle***. The structure and function of the heart muscle are covered in ***section 9.*** This section focuses mainly on the skeletal muscles - the muscles responsible for all voluntary and reflex movements.

Illustration 1: Anterior View. The Major Muscles of the Body

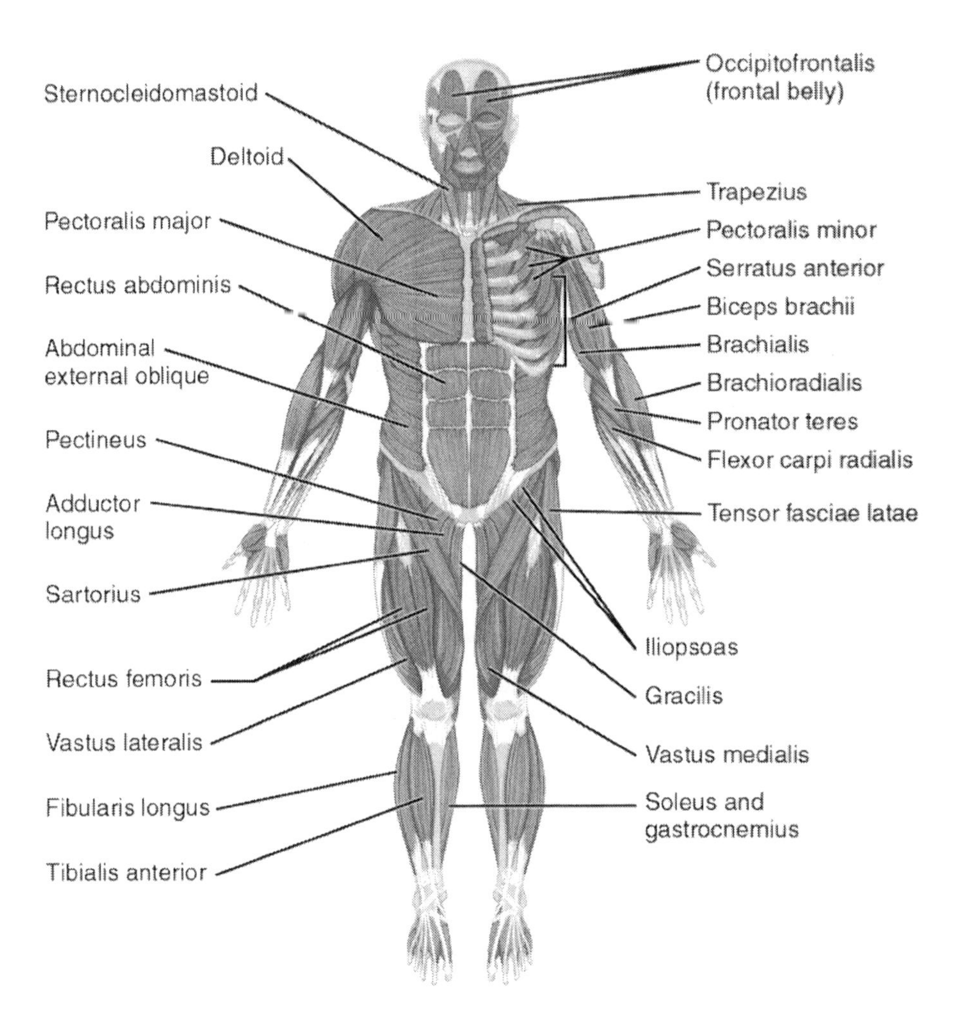

Illustration 2: Posterior View of the Major Muscles of the Body

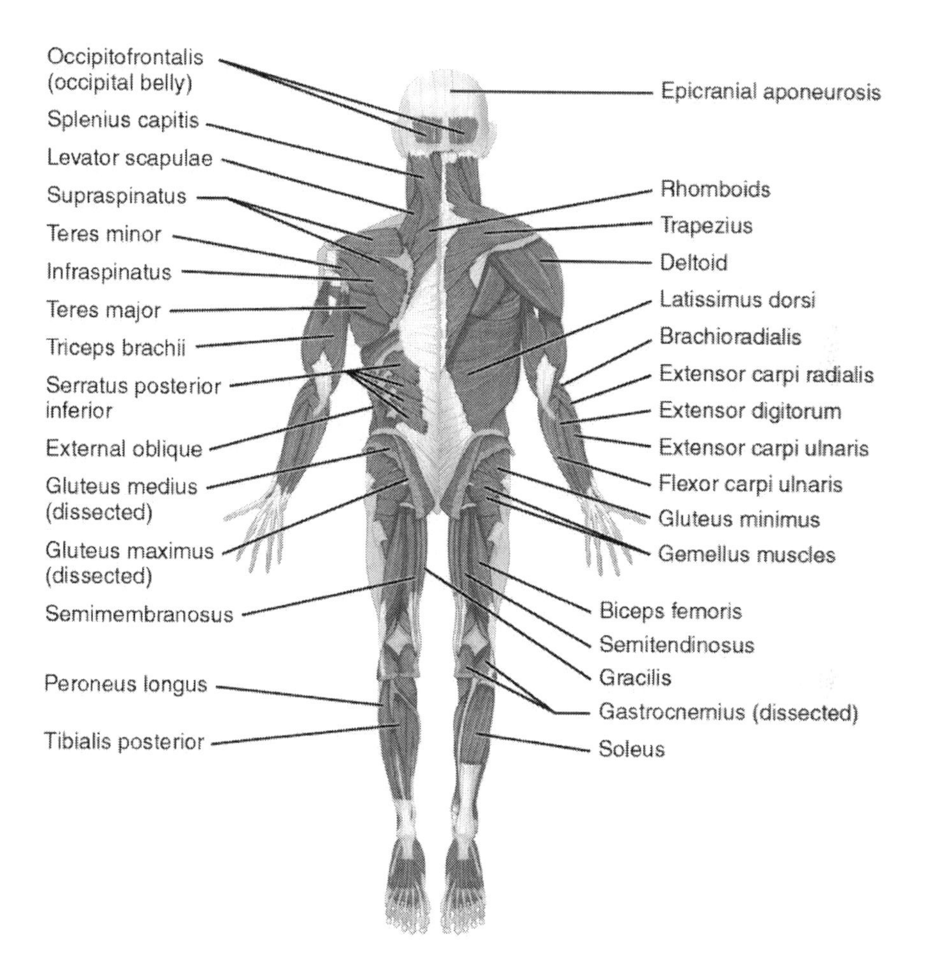

The Structure of a Skeletal Muscle

Skeletal muscles are made up of **muscle fibers**. Muscle fibers are large cells that contain multiple nuclei and internal fibrous structure. They are composed of many **myofibrils** - threadlike structures of protein filaments. These myofibrils contain even finer fibers called **actin** (thin filaments) and **myosin** (thick filaments).

Muscle fibers are covered by **endomysium** - a connective tissue layer that covers each individual skeletal muscle fiber. The skeletal muscle fiber is enclosed by a plasma membrane called **sarcolemma** which contains **sarcoplasm** (the cytoplasm of muscle cells).

Bundles of muscle fibers are called **fascicles**. Fascicles are surrounded by a sheath of connective tissue called **perimysium.** The perimysium binds the muscle fibers together into a fascicle. The fascicles are bound together by **epimysium (**a sheath of fibrous elastic tissue called) to form a muscle.

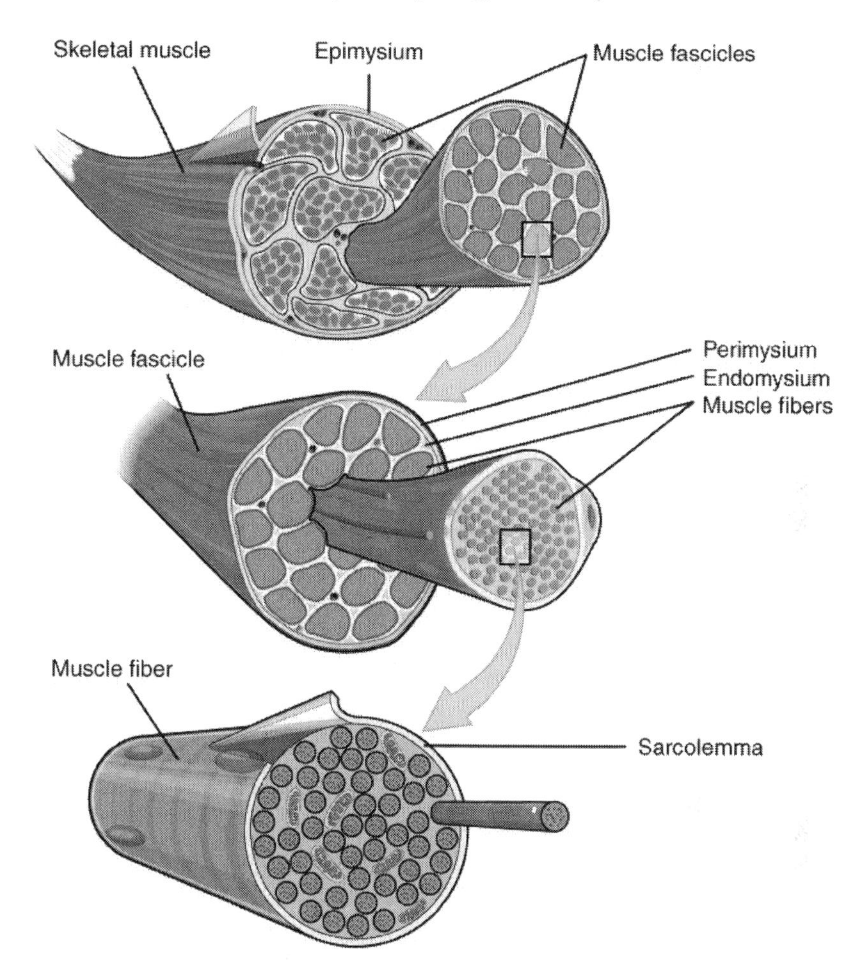

How Muscles Attach to Bones

Beyond the muscle, the epimysium becomes a tendon. Muscles attach to the bone either directly or indirectly. Attachment is considered to be 'direct' when the muscle's epimysium fuses with the periosteum (the connective tissue that envelops bones). Attachment is considered 'indirect' when the epimysium extends beyond the muscle to form a tendon which attaches to the bone. In the human body, indirect attachment is the most common way by which skeletal muscles attach to bones.

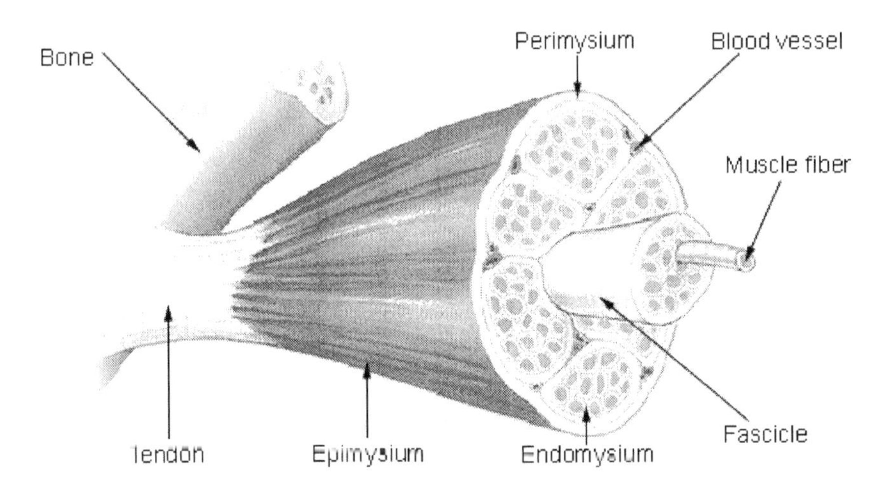

Tendons are strong fibrous connective tissue that attach to the periosteum (the connective tissue that wraps all bone except at the surface of the joints. Tendons should not be confused with ligaments. (**Ligaments** are strong fibrous connective tissue that connects two bones or holds together the joint).

Origin and Insertion

Most skeletal muscles attach to the bone through a tendon. During contraction, one of the bones to which the muscle is attached remains relatively stationary whilst the another bone (the more movable bone) is pulled in toward the 'stationary' or less movable bone.

- **The origin** is the attachment site where the muscle attaches to the 'stationary' bone. The origin is typically found on the proximal end of the bone (closer to the body relative to the insertion).
- **The insertion** is the attachment site where the muscle attaches to the more movable bone, the site that does move

when the muscle contracts. The insertion site is typically on the distal end.

Basic Muscle Movement

Muscles work together in groups because 'team work' enhances a particular movement. During a particular movement, depending on their origin or insertion, different muscles will play different roles. The muscle that does most of the moving is called an **'agonist'** or **'prime mover.'** Groups of skeletal muscles that contract at the same time to help move a body part are called '**synergists**.' Synergists are, in other words, muscles that help the prime mover achieve a certain body movement.

Muscles that act together to move a body part, but where one group of muscle counteracts another group of muscles, are said to be antagonistic. These muscles are also called **'antagonists.'**

The skeletal muscle can perform several types of movement.

1. **Flexion and Extension**

Flexion and extension are both movements that involve anterior or posterior movements of the body. Flexion is the bending (decreasing) of the joint angle. Extension is the straightening (increasing) of the joint angle.

The illustration below displays flexion and extension at the shoulder and knees (angular movements). These movements can also take place at the hip, elbow, wrist, and the joints of the hand and the feet. Flexion and extension movements take place in the sagittal plane of movement (***see Section 2***).

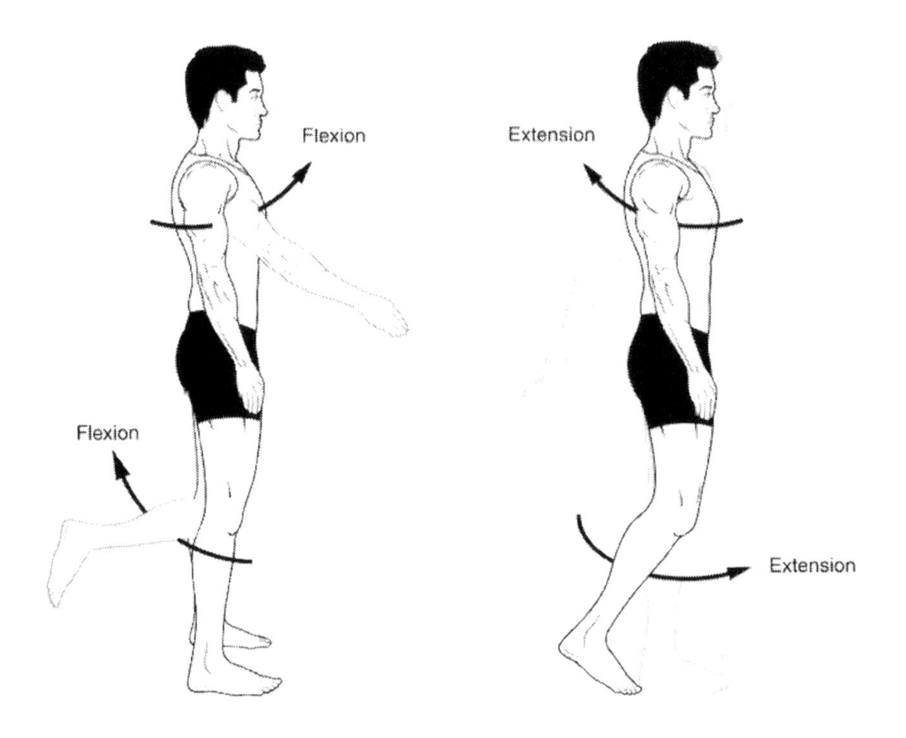

2. **Abduction, adduction, and circumduction**

Abduction entails 'moving away from the midline of the body.' Adduction involves 'moving toward the midline of the body.' Abduction and adduction movements take place in the coronal plane of motion.

Circumduction is the circular movement of a body part (e.g. the hand, fingers, or a limb). Circumduction uses a combination of flexion, extension, abduction and adduction movements.

The illustration below displays abduction, adduction and circumduction of the upper limb at the shoulder (angular movements). Abduction and adduction are movements of the limbs, fingers, hands, and toes. Abduction, adduction and circumduction can all take place at the hip, shoulder, hip, MCP and MTP joint.

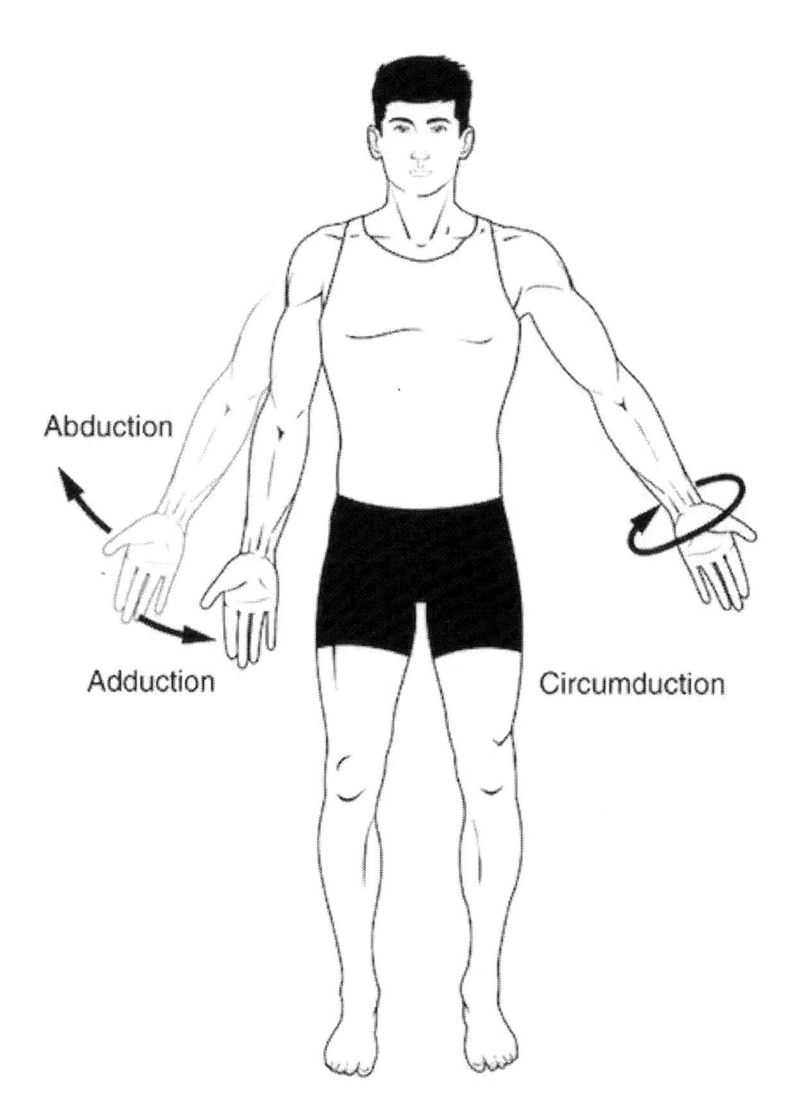

3. Medial and Lateral Rotation

Medial or internal rotation involves 'turning toward the midline of the body'. Lateral or external rotation involves 'turning away from the midline of the body.'

The illustration below displays rotation of the head, neck, and the lower limb. Medial rotation of the upper limb of the should for example, involves turning the limb's anterior surface toward the body's midline. Lateral rotation of the upper limb of the shoulder in turn, involves turning the limb away from the midline of the body.

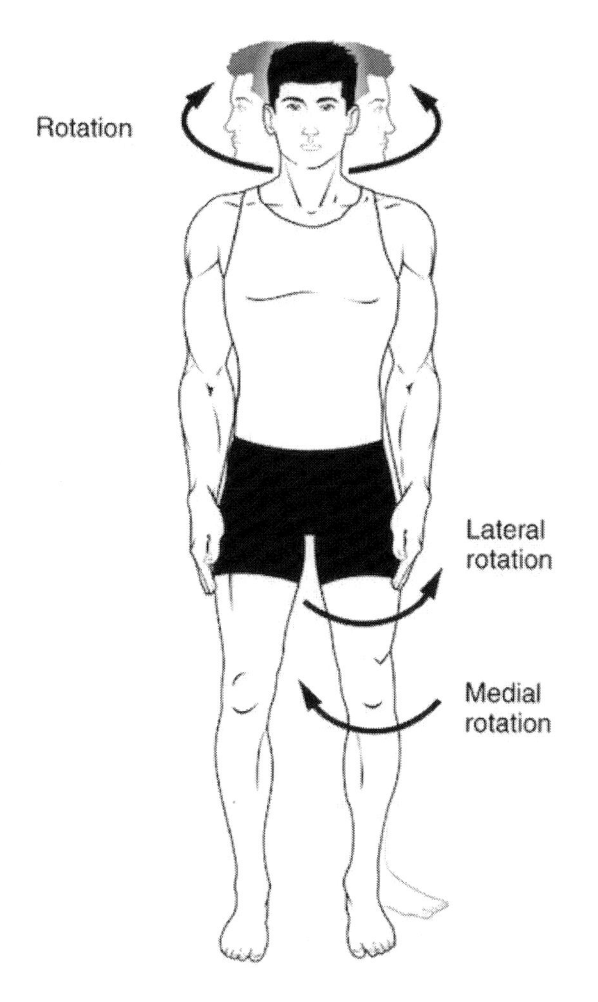

4. **Protraction and Retraction**

Protraction is 'moving forward' and retraction is 'moving backward.' Protraction of the mandible (jawbone) for example, pushes the chin forward, whereas retraction of the mandible pulls the chin inwards.

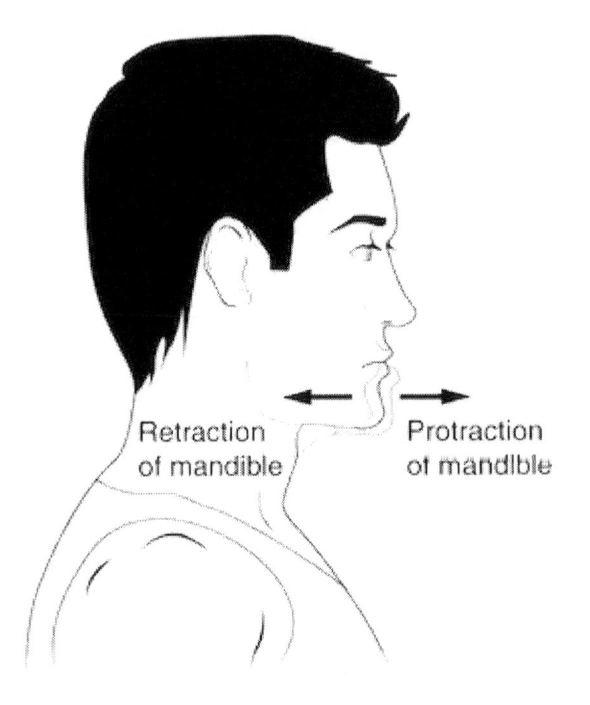

5. **Pronation and Supination**

Pronation is 'turning downward' and supination is 'turning upward.' Supination of the forearm, as illustrated in the diagram below (S), turns the palm of the hand into a forward position. Pronation of the forearm (P), turns the hand to the palm into a backward position.

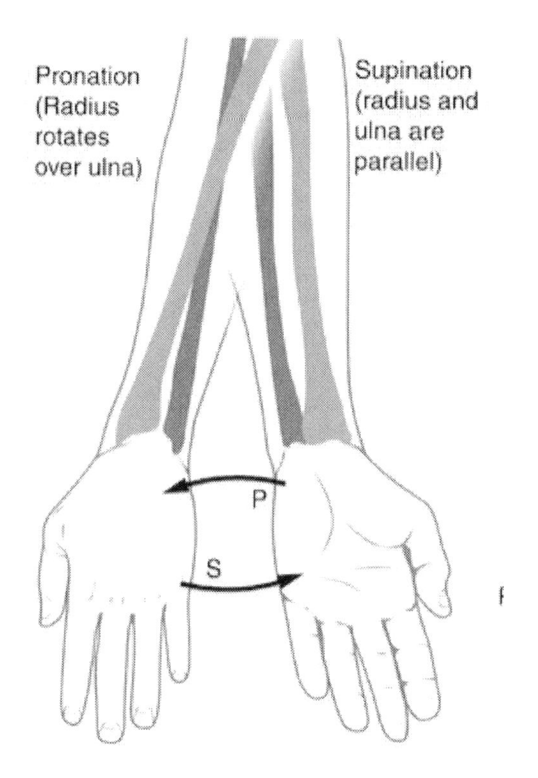

6. **Eversion and Inversion**

Eversion is 'turning outward' and inversion is 'turning inward.' Eversion of the foot for example, turns the sole of the foot away from the body's midline, whereas inversion turns the bottom of the foot toward the body's midline.

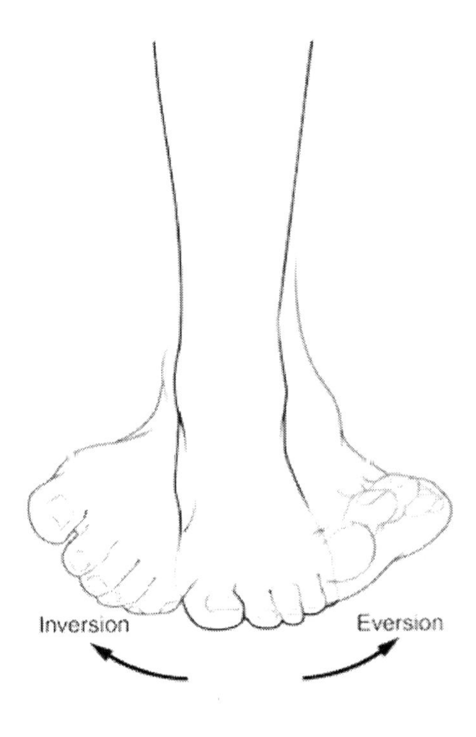
Inversion
Eversion

The Skeletal System

The human skeleton is made up of 260 bones. 80 bones are found in the **axial skeleton** and 125 bones are found in the **appendicular skeleton**. The axial skeleton consists of the bones which are found along the midline of the human body and the appendicular skeleton contains the bones and joints of the appendages (the upper and lower limbs) and the two girdles. The body has two girdles. The **pectoral girdle**, also called the shoulder girdle, encircles the top vertebral column. The **pelvic girdle** encircles the bottom vertebral column.

The Axial Skeleton forms the 'central axis' of the human body and includes the following bones:

- Bones of the skull (facial and cranial bones)
- Ossicles (small bone found in the middle ear)
- Hyoid bone (of the throat)
- Vertebrae
- Rib cage (also called thoracic cage)
- Sternum

The Appendicular skeleton consists of the girdles (which join the appendages to the axial skeleton) and the bones of the limbs which are found in six major regions:

- Pectoral girdles (clavicle and scapula)
- Arms and forearms (humerus, ulna, and radius)
- Hands (left and right carpals, metacarpals, proximal phalanges intermediate phalanges, and distal phalanges)
- Pelvis (left and right hip bone)
- Thighs and legs (left and right femur, patella, tibia, and fibula)

- Feet and ankles (left and right tarsals, metatarsals, proximal phalanges, intermediate phalanges, and distal phalanges)

Anterior View of the Human Skeleton

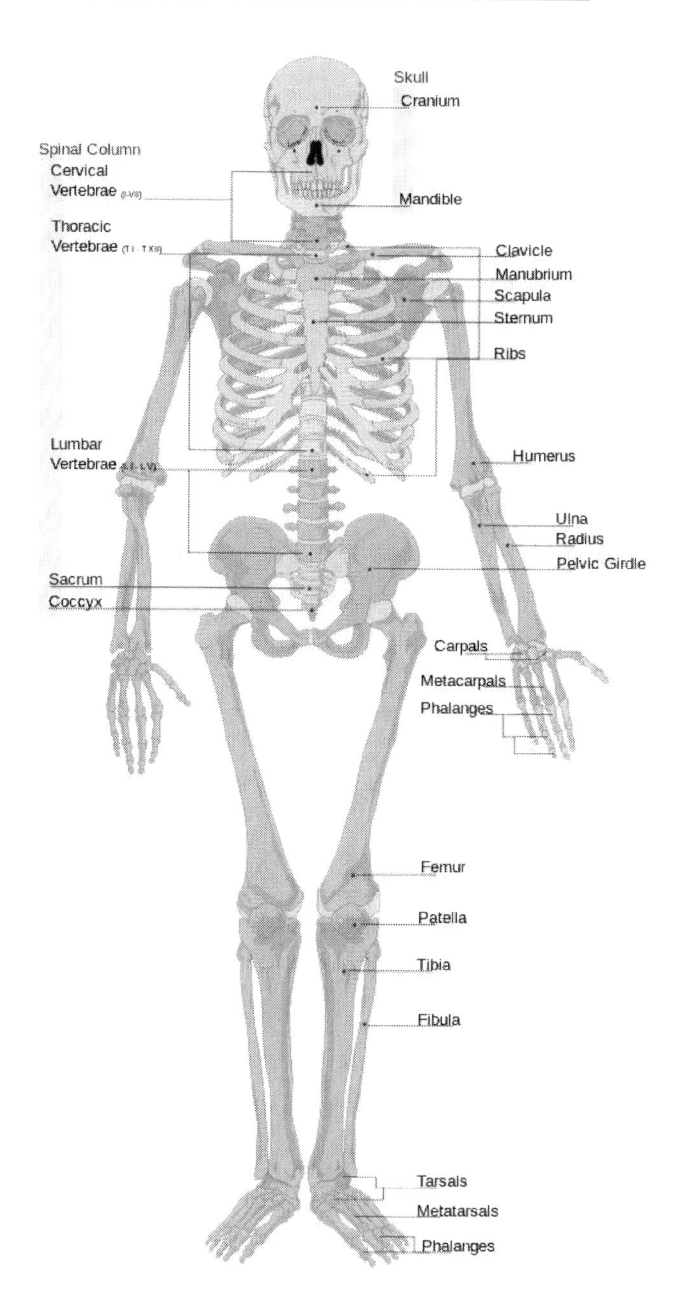

Posterior View of the Human Skeleton

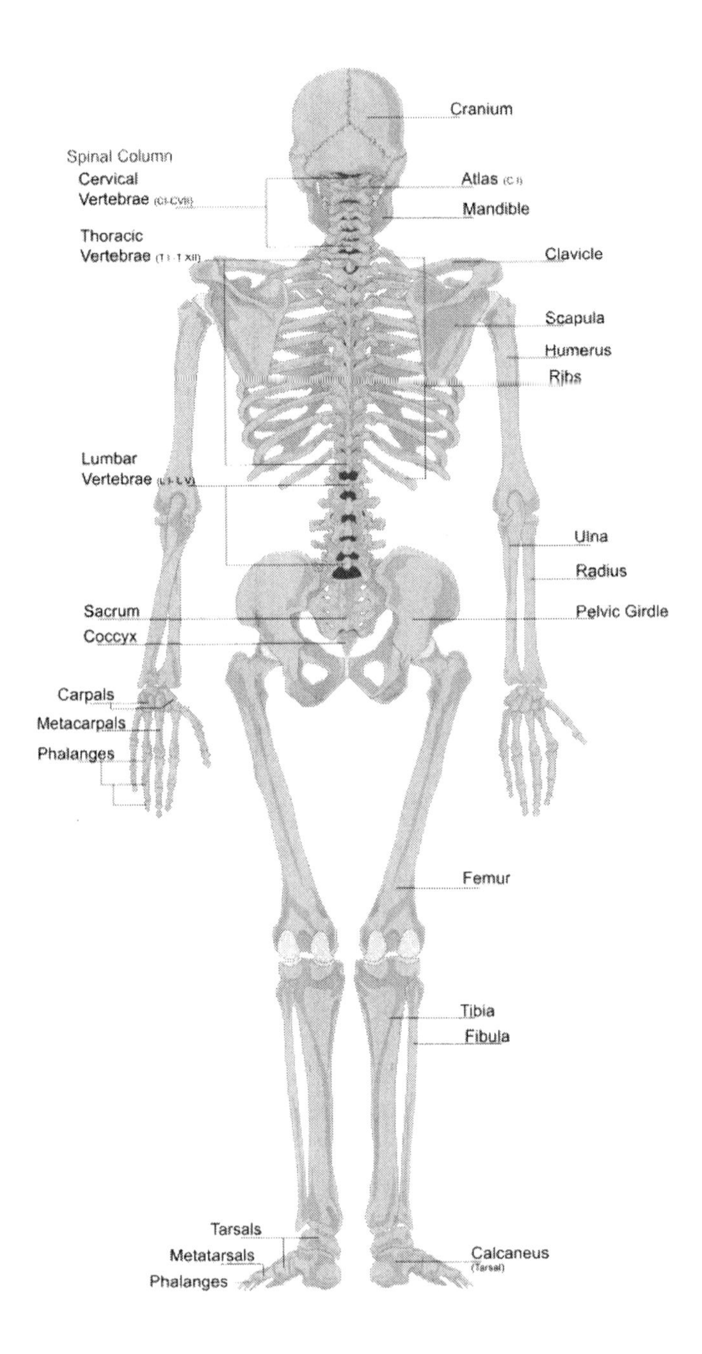

Bones

Bones perform various functions. Such functions include: stabilizing and supporting the body, protecting internal structures and vital organs, storing salts, producing red blood cells (in the bone marrow), and providing an attachment site for muscles, tendons, and ligaments. Bones are classified by shape, as being either long, short, flat, irregular (e.g. the mandible and the vertebrae), sutural (irregular bones located along the sutures of the skull), or sesamoid (a small independent bone developed in a tendon).

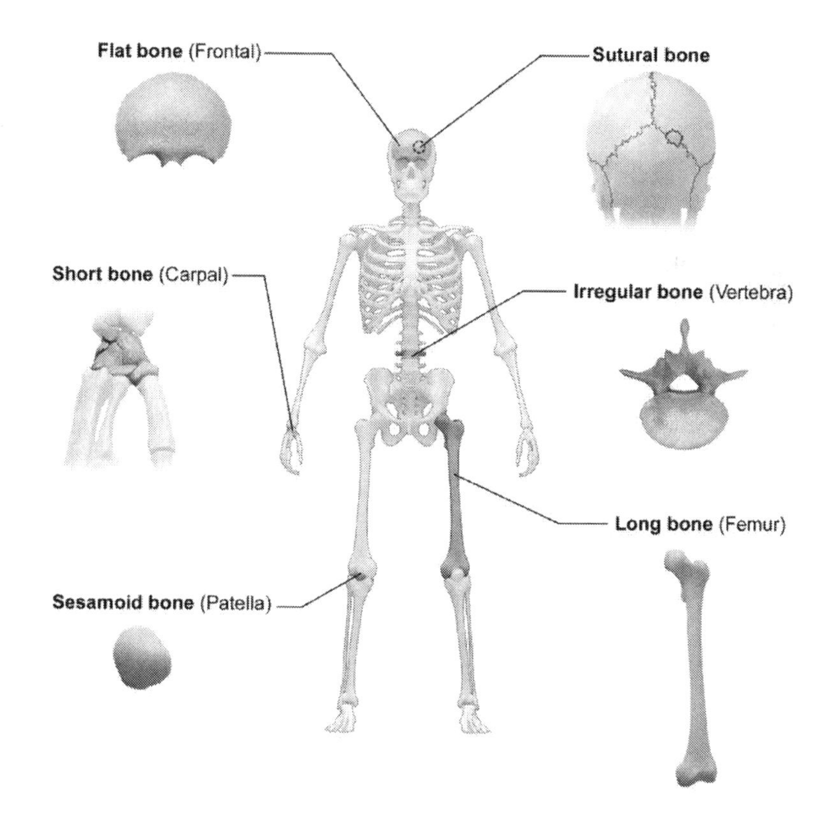

Bone Tissue and Structure

Each bone is made up of an outer layer of compact bone that is dense and smooth and an inner layer of spongy bone.

- **Cortical (compact) bone** (the outer layer) is composed of a hard calcified matrix that contains a dense layer of bone cells.
  Compact bone tissue also contains **lamellae** (bone layers) and **haversian canals** (central canals). Blood reaches the bones through either the haversian canals, the Volkmann's canals, or through vessels found in the bone marrow. **Lacunae** are small cavities within the bone matrix, which contain **osteocytes** (bone cells). Small canals called **canaliculi** connect the lacunae and also provide nutrients to the bone.

- **Spongy (cancellous) bone** (the middle layer) is composed of **trabeculae,** which are plates of bone tissue that form cancellous bone. In some bones, red bone marrow fills the spaces between the trabeculae.

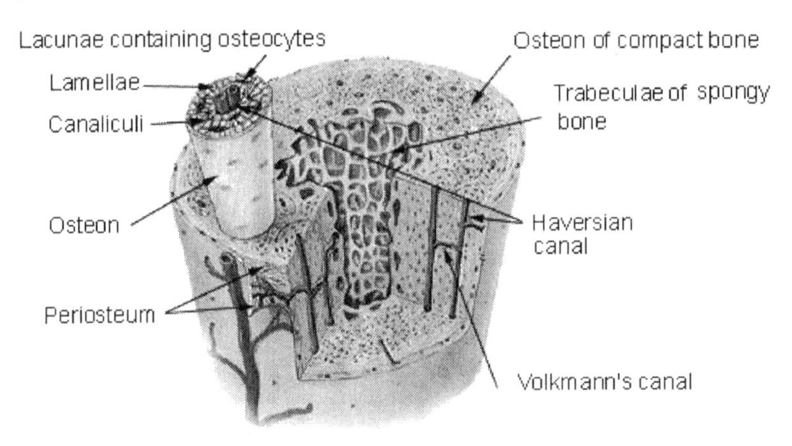

Long Bone

The long bone consists of three main parts: ***the diaphysis (the shaft), epiphysis (the ends), and the metaphysis***, this is where the diaphysis merges with the epiphysis. Examples of long bones include the thighbone (*femur*) and the forearm bone (*radius*).

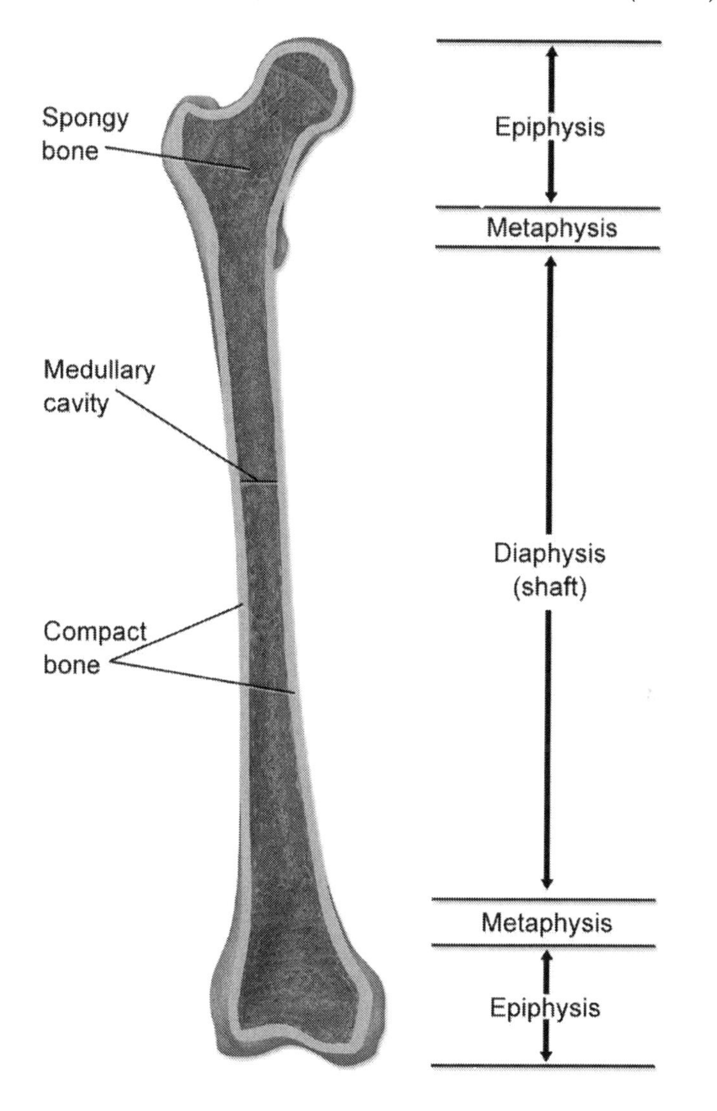

Bone Development: Growth and Remodeling

Ossification is the natural process of bone formation. Bones starts off in the form of cartilage which is then replaced by bone - a process called ossification. There are two types of ossification:

intramembranous and *endochondral.*

Intramembranous Ossification

Intramembranous ossification is the replacement of connective tissue membranes with bony tissue. It is one of the two processes during fetal development of the mammalian skeletal system by which bone tissue is created. Bones that are formed in this manner, which includes certain irregular bones and some flat bones of the skull, are called intramembranous bones.

Endochondral Ossification

Endochondral ossification is the replacement of hyaline cartilage with bony tissue. The majority of bones in the human body, called endochondral bones, are formed through endochondral ossification. Endochondral ossification also plays a central role in the formation and growth of long bones, and in the healing or remodeling of bone fractures.

Osteogenesis

Osteogenesis is the formation of bone (the ossification of cartilage into bone). Both terms, osteogenesis and ossification are typically used synonymously to indicate the process of bone formation. In children, osteogenesis starts at the ninth week of fetal development. Ossification and bone growth occur at the same time in children: the cartilage cells divide, the bone lengthens, and the cartilage calcifies. Once adult height is reached, the cartilage cells of the bone stop to divide, which also puts an end on bone growth. Bone development in adults however, continues in the form of fracture repair and remodeling (to meet changing lifestyles).

Bone Growth

The three cells which are involved in bone growth, development and remodeling are osteoblasts (bone-forming cells), osteocytes (mature bone cells), and osteoclasts (cells that break down and reabsorb bone).

In a long bone, the epiphyseal plate is the site of bone growth. In immature bones, ossification occurs in the layer of hyaline cartilage. Cartilage is formed on the epiphyseal plate by mitosis. As chondrocytes in the diaphysis degenerate, osteoblasts migrate to the region and begin to ossify the cartilage into bone tissue. When osteoblasts reach the membrane, they deposit bony matrix around themselves at which point they are called osteocytes (bone cells). This process, which allows the diaphysis to grow in length, continues until cartilage growth slows and comes to an end once adult height is reached. When cartilage growth stops, the epiphyseal plate becomes completed ossified, and longitudinal growth of bone ceases.

Although bone stops growing in length at this stage, they can still continue to grow in diameter or thickness throughout life in response to lifestyle changes, e.g. due to increased weight or muscle activity.

Bone Remodeling

Bone remodeling or bone metabolism, is the replacement of old bone with new bone. Bone remodeling continues even once adult height is attained. Ordinary activity causes microscopic cracks to form in the bone. These are then dissolved and replaced with new bone tissue. The process of bone remodeling entails removal of mature bone tissue from the skeleton (called **bone resorption** – which occurs when osteoclasts break down bone to release minerals), followed by bone formation (**ossification**).

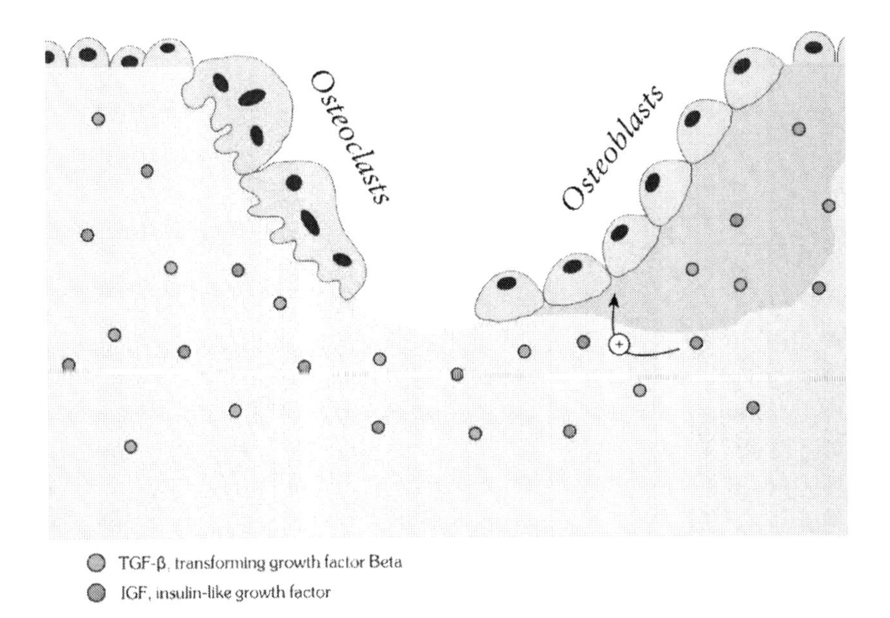

Joints

Joints are the contact point at which bones are held together. Joints can be classified by either their function (i.e. by their range of motion) or their structure. The human body contains three major types of joints that are classified according to their function and three types of joints which are classified pursuant to their structure.

Classification by Function:

1. **Diathrosis:** These are freely movable joints which are also called synovial joints. Diarthroses, e.g. the elbows, are joined together by ligaments.
2. **Amphiarthrosis:** These are joints that are slightly movable and which are connected by either hyaline cartilage or

fibrocartilage. The intervertebral disks for example are amphiarthroses.

3. **Synarthrosis:** These are joints that are immovable, e.g. joints between the bones of the skull. Synarthroses are joints together by sutures (a layer of fibrous connective tissue).

Classification by Structure:

1. **Synovial joints are also called diarthroses.**
2. **Cartilaginous joints are also called amphiarthroses.**
3. **Fibrous joints:** Fibrous joints allow for little movement and are joined together by fibrous connective tissue. Examples include sutures, the dental alveolar joint, and the radioulnar joints (joint between the two bones in the forearm).

Cartilage

Cartilage is a firm connective tissue that consists mostly of protein fibers. Cartilage is the main component of joints and also supports other structures in the body, including the larynx, the intervertebral disks and the auditory canal. Compared to bone tissue, cartilage has fewer cells and little to no blood supply, which affects its ability to heal. There are three types of cartilage - hyaline, fibrous, and elastic cartilage.

- **Hyaline Cartilage:** the most common cartilage type found in the body and a major component of synovial joints and articular bone surfaces. Hyaline cartilage is also found in the nasal septum, the bronchi, and the trachea.
- **Fibrous Cartilage or Fibrocartilage**: a strong, fibrous and spongy tissue that forms the pubis symphysis.
- **Elastic Cartilage**: abundant in elastic fibers, it is the most flexible cartilage in the human body. Elastic cartilage can be found in the epiglottis, the auditory canal, and in the external ear.

Synovial Joints

Synovial joints are freely movable and include most of the joints found in the arms and legs. Structures found in the synovial joint include the following:

- **Bursae:** Bursae are saclike cavities which contain fluid. They are found around joints at friction points where they act to facilitate the gliding of tendons and muscles over surfaces that are either bony or ligamentous. Most bursae are found at the hip, knee, elbow, and shoulder.
- **Joint cavity or synovial cavity:** The joint cavity is a space that separates the articulating surfaces of two bones.
- **Joint capsule or articular capsule**: This is the envelope that surrounds the synovial joint comprising of an outer fibrous membrane and an inner synovial membrane.
- **Ligaments**: Ligaments are strong fibrous connective tissue that connect bones at the joints and which strengthen the capsule.

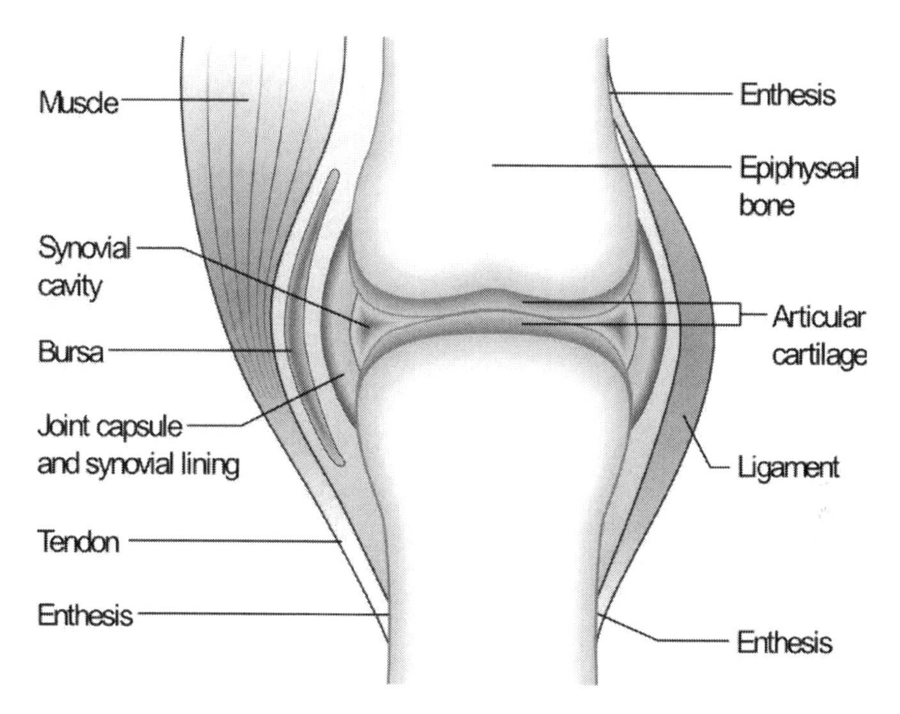

Based on their function and structure, synovial joints can be further classified into subcategories: pivot (a), hinge (b), saddle (c), plane (d), condylar (e), and ball-and-socket (f).

- **Pivot joints** are found where a rounded bone portion fits into the grove portion of another bone.
- **Hinge joints** are found where a convex shaped bone portion fits into a concave shaped bone portion.
- **Saddle joints** have a saddle-shaped surface. The only saddle joints found in the human body are the carpometacarpal joints of the thumb.
- **Plane joints** are joints that only allow gliding (or sliding) movements. In plane joints, a convex shaped bone portion fits into a concave shaped bone portion.
- **Condylar joints** are found where an oval shaped bone portion fits into a concavity of another bone.

- **Ball-and-socket joints** are found where a spherical head of a bone fits into a concave shaped portion of another bone.

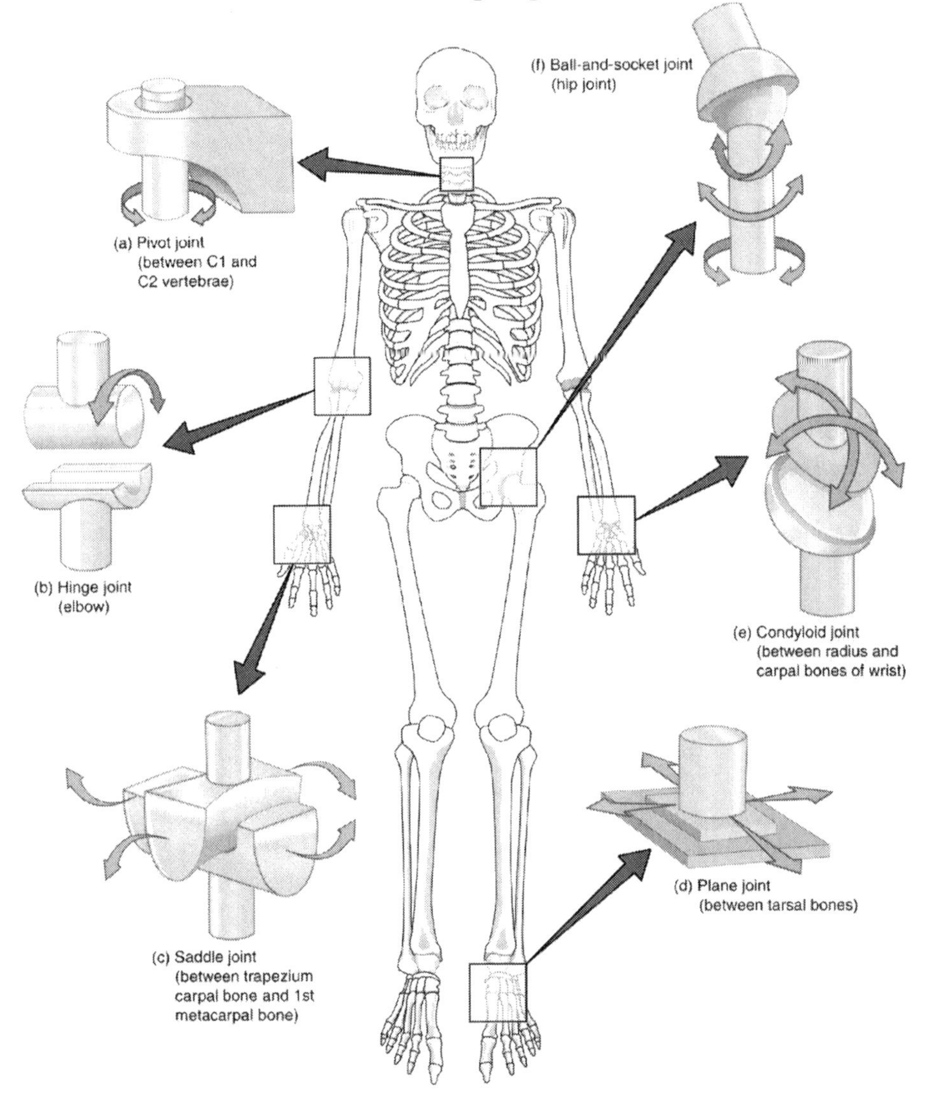

Section 6: The Neurosensory System and the Sense Organs

The nervous system coordinates all bodily functions, allowing the body to respond to internal and external stimuli. The sensory system is the part of the nervous system that is responsible for relaying sensory information. In this chapter, we will cover the anatomy and physiology of the nervous system, as well as special sense organs of the body and their relative functions.

The Nervous System

The nervous system has two main types of cell: conducting cells calls **neurons** and supportive cells called **neuroglia.**

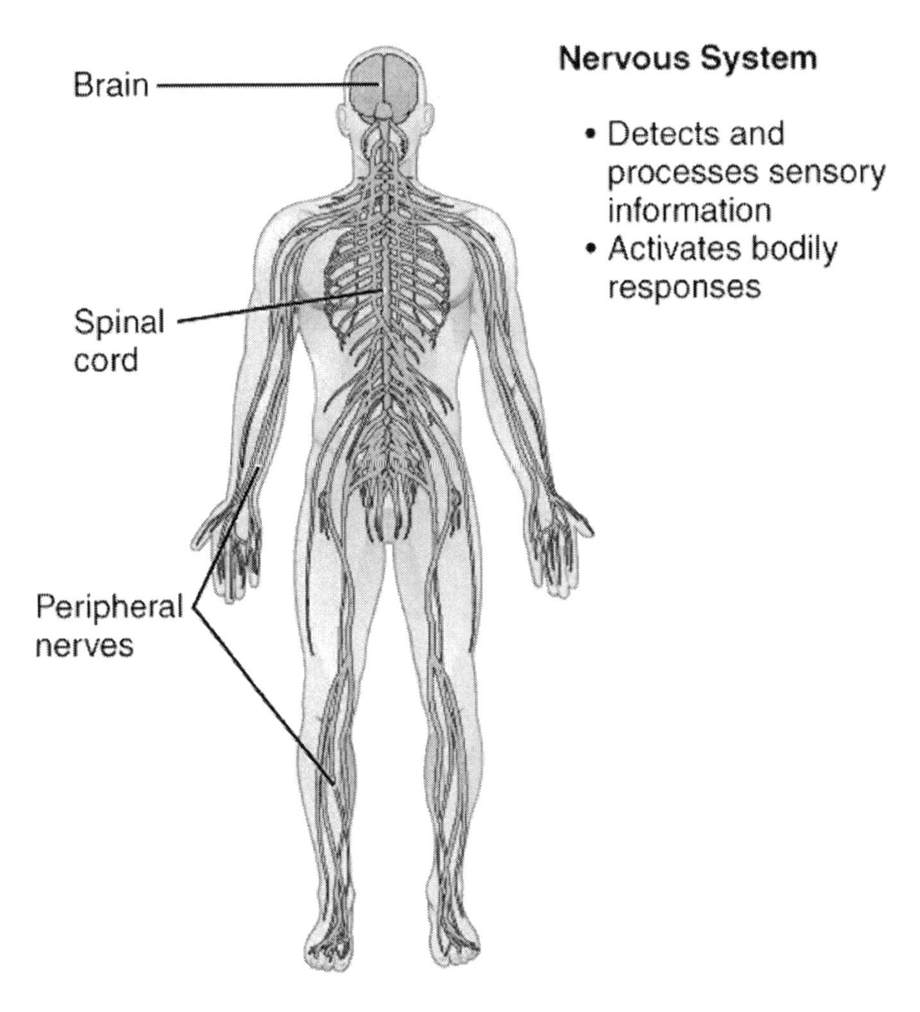

Neuron Structure

A neuron comprises of one axon and multiple dendrites. The **axon**, also called the **nerve fiber**, is the part of the neuron responsible for the transmission of information to different neurons or other body structures. Axons carry nerve impulses away from the cell body. **Dendrites** are short branched extensions of the neuron which conduct impulses toward the cell body, i.e. they responsible for receiving nerve impulses from other cells.

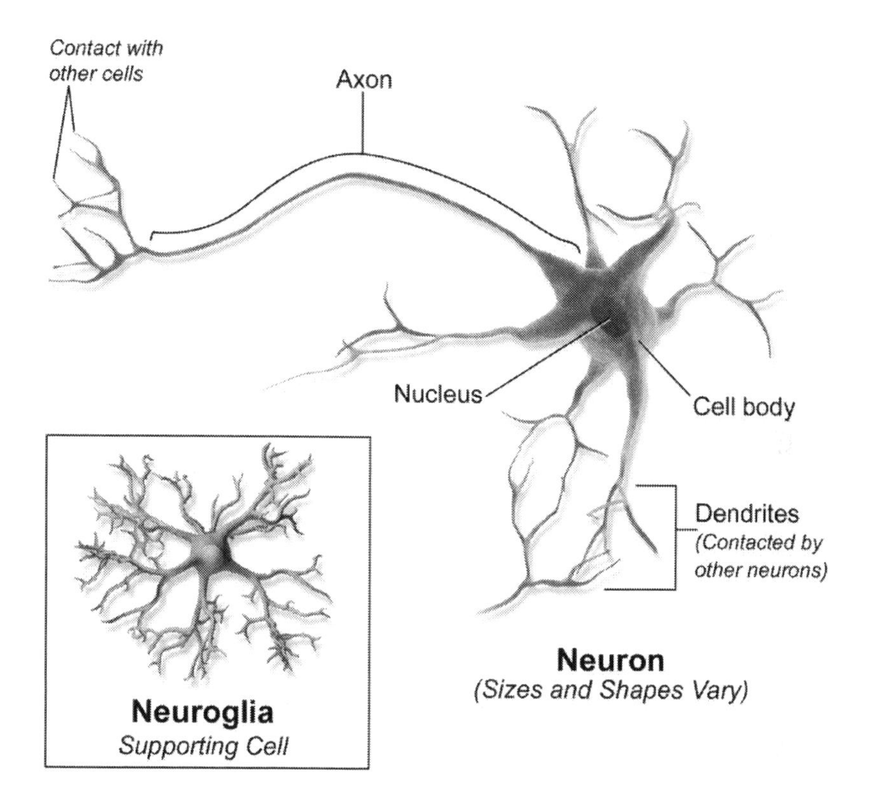

Neuroglia

Neuroglia, also called glial cells, are the supportive cells of the nervous system. Four types of neuroglia exist in the nervous system.

1. **Ependymal cells** help produce cerebrospinal fluid (CSF). Ependymal cells line the central canal of the spinal cord as well as the ventricular system of the brain.
2. **Oligodendrocytes** are glial cells that create the myelin sheath in the central nervous system.
3. **Astrocytes**, also called astroglia, star-shaped glial cells. They are present throughout the nervous system, where

they supply nutrients to the neurons and also help neurons maintain their electrical potential. They also perform other functions, such as supporting endothelial cells which form the blood-brain barrier which prevents harmful molecules from entering the brain.

4. **Microglia** function as phagocytic cells in the central nervous system where they protect the body by engulfing and ingesting microorganisms and waste products from injured neurons.

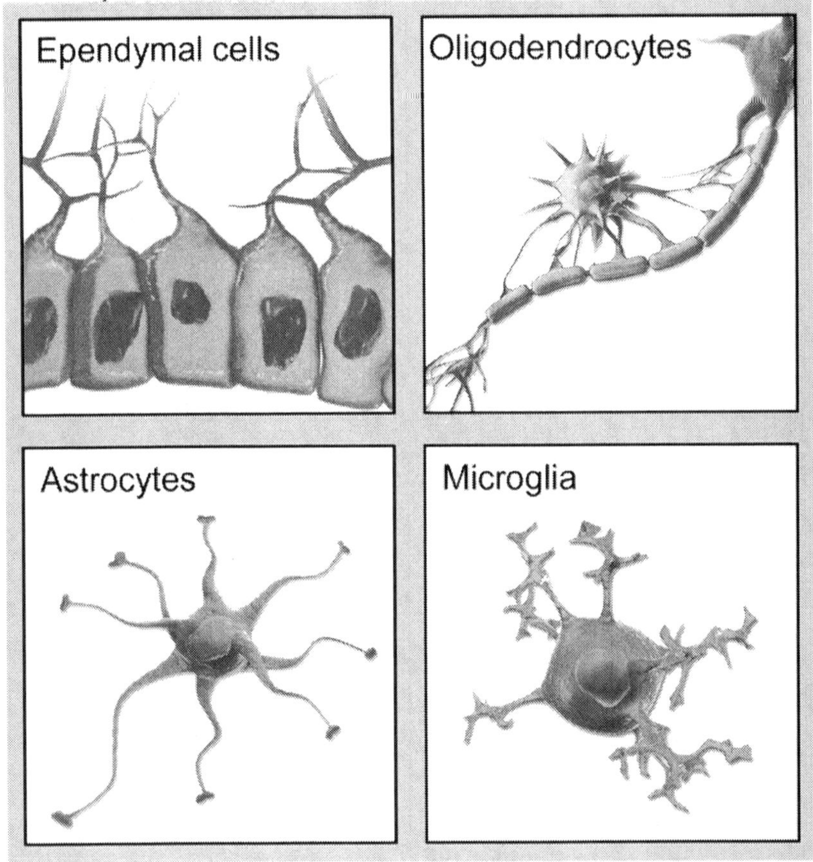

Neurotransmission

Neurons (nerve cells) are responsible for **neurotransmission (also called synaptic transmission)**. Neurotransmission, a process essential for communication between two neurons, is the conduction of nerve impulses through the body. During neurotransmission, neurons release chemical messengers called **neurotransmitters.** Neurotransmitters 'transmit' signals from one neuron to another neuron.

Neurons can receive (via the dendrites) and transmit (through the axon) electrochemical messages. Dendrites receive impulses sent from other cells and conduct these impulses toward the cell body. Axons conduct impulses away from the cell body.

Neuron activity can be stimulated by any of the following:

- **A chemical stimulus**, e.g. a chemical released by the body
- **A mechanical stimulus**, e.g. pressure or 'touch'
- **A thermal stimulus**, e.g. heat or cold

How Neurotransmissions Work

1. Once the impulse has been generated, it travels along the axon of the neuron.
2. As it reaches the end of the axon, the synaptic vesicles in the presynaptic nerve terminal are stimulated.
3. This provokes the release of chemical substances (neurotransmitter molecules - also called neurotransmitters) into the synaptic cleft between the two neurons (the synapse).
4. The neurotransmitters then diffuse across the synapse to the receiving neuron. The receptor sites are located on the dendrites of the receiving neuron.
5. After crossing the synaptic cleft, the neurotransmitters bind to receptors on the postsynaptic membrane. This inhibits or promotes stimulation of the postsynaptic neuron.

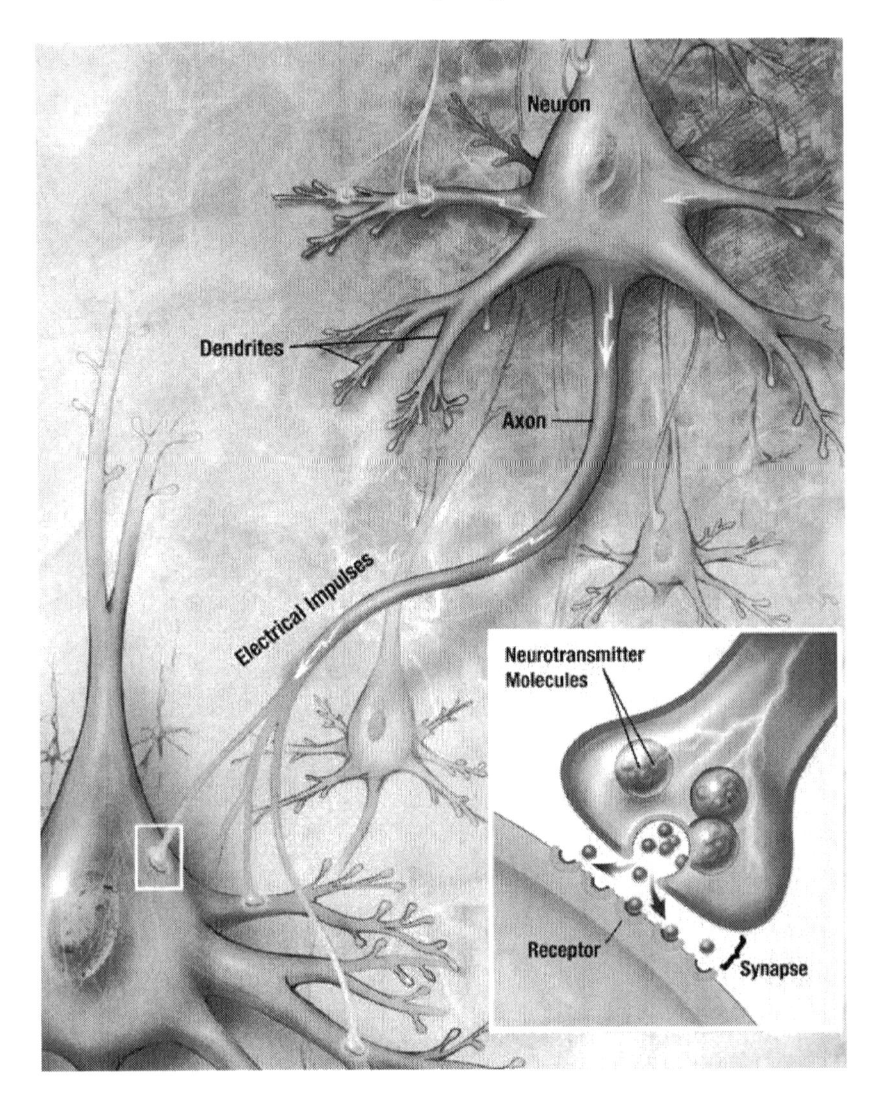

The Reflex Arc

The reflex arc is the nerve pathway involved in the transmission of a sensory impulse to a motor neuron. The reflex arc requires a sensory neuron one side (**afferent**) and a motor neuron on the other (**efferent**). Sensory neurons are neurons that transmit sensory information such as sight or sound. Muscle neurons, also called motoneurons, are located in the spinal cord. Their fibers control effector organs, such as muscles and glands.

A stimulus triggers a sensory impulse which then passes through the dorsal root to the spinal cord. Two synaptic transmission occur at the same time. One impulse travels along the sensory neuron to the brain and another transmits it directly, via a relay neuron, to a motor neuron. The muscle neuron in turn transmits the impulse to a muscle or a gland, producing an immediate action or movement. A reflex is an action or movement that is not controlled consciously.

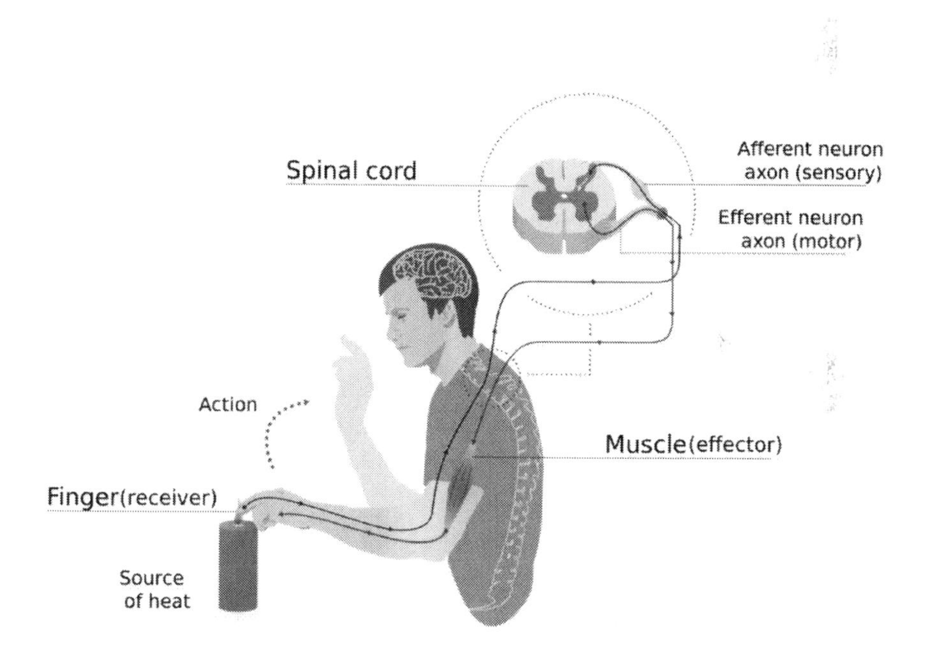

The Central Nervous System (CNS)

The CNS controls most of the functions of the body and the mind and comprises the brain and the spinal cord.

The Brain

The brain includes the ***cerebrum*** (the forebrain), ***cerebellum*** (the hindbrain), ***brain stem, diencephalon*** (the thalamus and hypothalamus), ***reticular activating system***, and the ***limbic system***

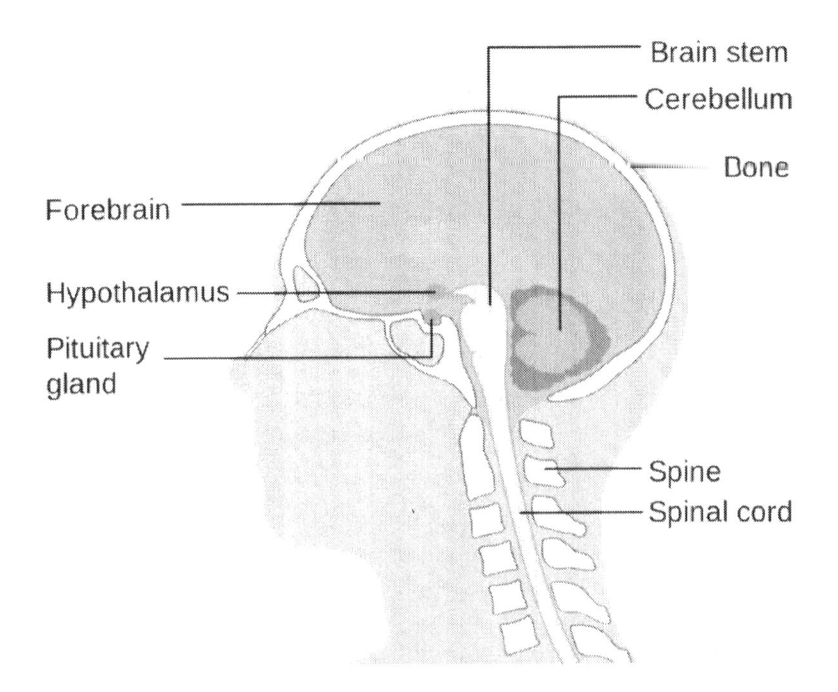

<u>Brain Structure and Function</u>

1. The Cerebrum

The cerebrum, also called forebrain, the principal part of the brain which is located in the front skull area. It contains the nerve center responsible for controlling sensory and motor activities and intelligence. The cerebrum (with the assistance of the cerebellum) controls all of the body's voluntary actions and movements.

Outer and Inner Layer

The cerebrum's outer layer is called the **cerebral cortex.** It is composed of folded grey matter (unmyelinated nerve fibers), which play an essential role in consciousness. The inner layer of the cerebrum is composed of white matter (myelinated nerve fibers) that contain **basal ganglia.** Basal ganglia are a group of structures which are involved in coordination and movement.

Cerebral Hemispheres

The cerebrum is made up of two cerebral hemispheres: the **right and left hemisphere**. The right hemisphere controls and processes impulse from the left side of the body and the left hemisphere controls and processes those received from the right side of the body. Communication between corresponding centers in each hemisphere is transmitted through a broad band of nerve fibers (called **corpus callosum**), which join the two hemispheres.

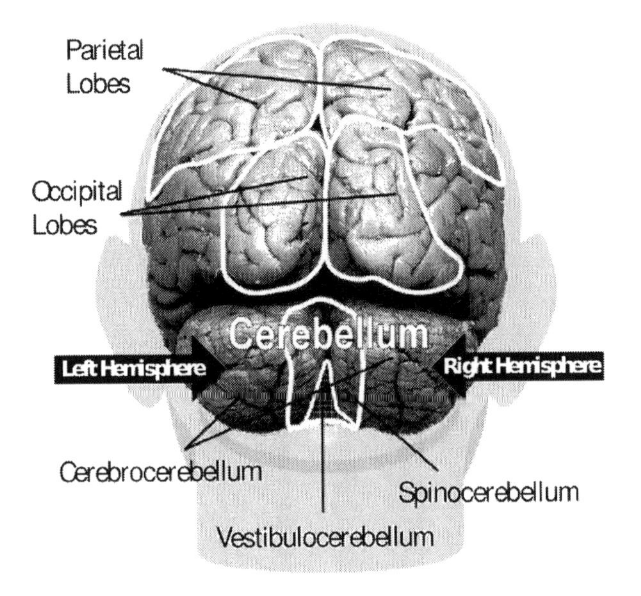

The cerebral hemispheres are each divided into four lobes: **the frontal, temporal, parietal, and occipital lobes.**

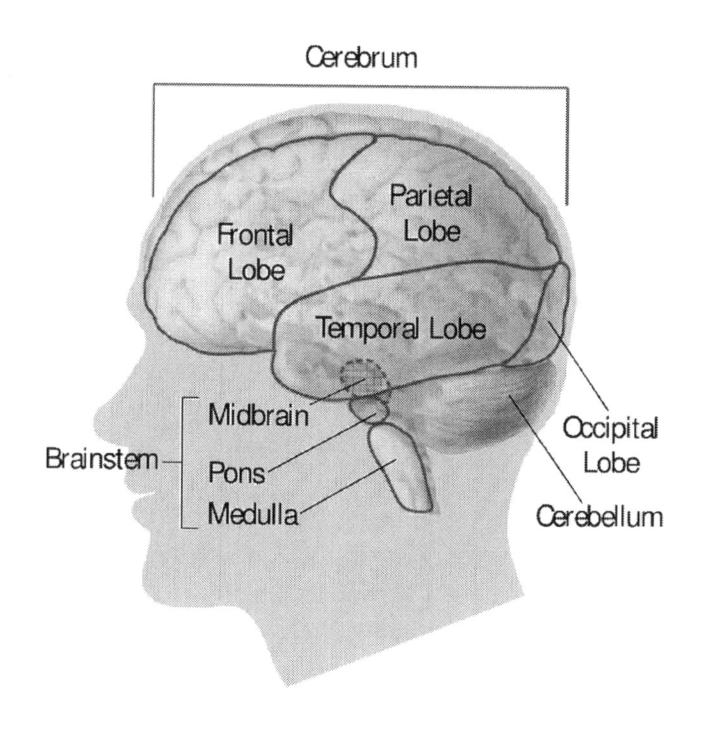

2. Cerebellum

The cerebellum is the second largest region of the brain. the cerebellum controls balance, coordinates muscle movement, and helps maintain muscle tone. The cerebellum is also divided into a left and a right hemisphere and also contains an outer cortex of grey matter and an inner core which contains white matter.

3. Brain Stem

The brain stem lies below the cerebrum and continues downward to form the spinal cord. The brain stem performs many basic functions, which includes the regulation of the heart rate, sleeping, breathing and eating. It also provides the pathway for nerve fibers between the lower and higher neural centers and supplies that majority of motor and sensory nerves to the neck and the face (via cranial nerves). In fact, ten out of the twelve pairs of cranial nerves originate from the brain stem.

It consists of the medulla oblongata, pons, and midbrain:

- **The Medulla Oblongata** forms the lowest part of the brainstem, i.e. is the continuation of the spinal cord within the skull. It controls autonomic (involuntary) functions such as vomiting, coughing, and hiccups. It also helps regulate heart and blood vessel function. breathing, digestion, sneezing, and swallowing.
- **The Pons** (which sit above the medulla) serves as a message station between several parts of the brain - notably, it helps relay messages from the cerebrum and the cerebellum. The pons also helps regulate and mediate sleep, respirations, swallowing, hearing, equilibrium, and taste.
- **The Midbrain** controls many important functions such as eye movement, pupillary reflexes and other functions relating t the visual and auditory systems. Some portions of the midbrain are also involved in the control of body movement.

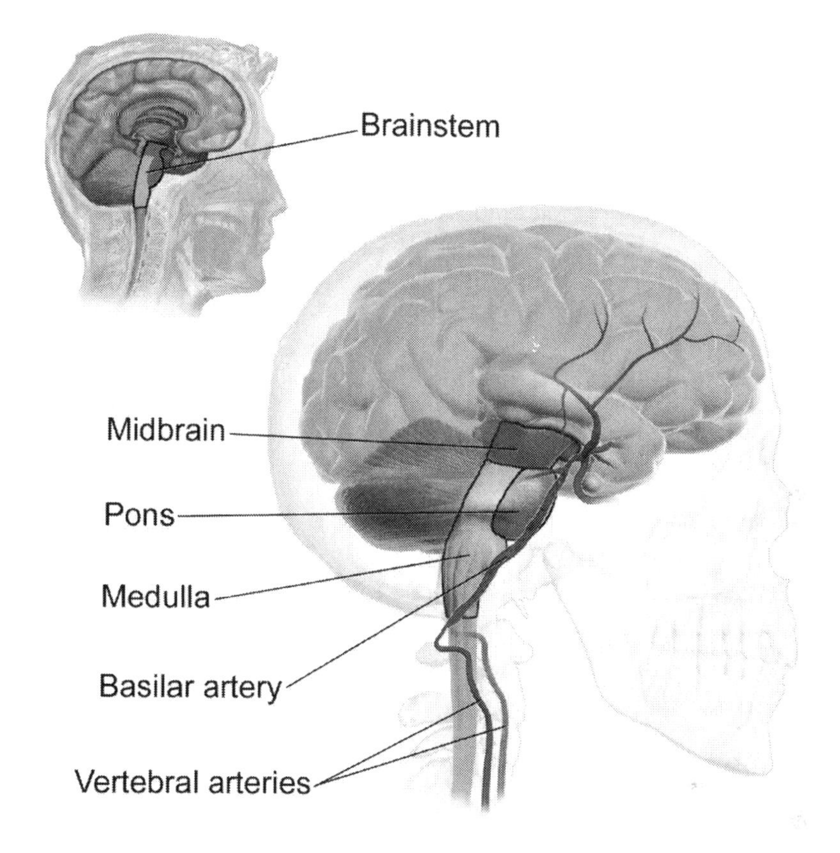

4. Diencephalon

The diencephalon, located between the cerebrum and the midbrain, is composed of the thalamus and the hypothalamus.

- **The Thalamus** is relays most sensory information (apart from those relating to the sense of smell) to the cerebral cortex. The thalamus acts as a center for sensory interpretation, including pain perception.
- **The Hypothalamus** lies below the thalamus and controls body temperature, thirst, hunger, as well as other sleep and emotional activity.

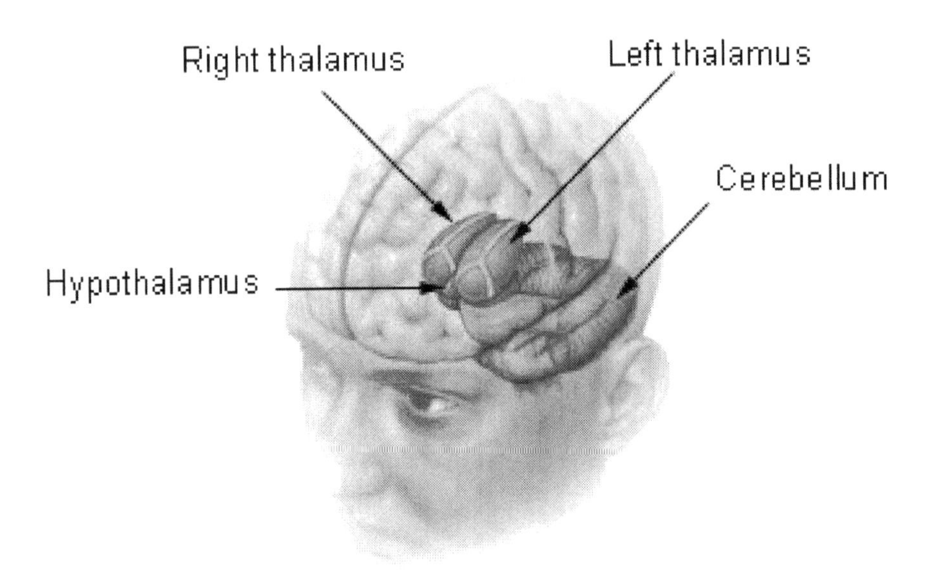

1. **The Limbic System**

The limbic system is a system of nerves and networks inside the brain which controls basic emotions (such as pleasure and fear) and drives (such as hunger and sexual arousal).

2. **The Reticular Activating System (RAS)**

The RAS is a diffuse network of nerve pathways that arouse and alert the cerebral cortex. The RAS is connected to the cerebrum, the cerebellum and the spinal cord and plays a central role in mediating consciousness and

The Spinal Cord

The spinal cord is the cord of nerve fibers and tissue that lies in the vertebral canal and which is connected to the brain. It connects almost all body parts to the brain, with which it forms the central nervous system. Spinal nerves, which carry autonomic, motor, and sensory signals, arise from the cord.

The cross section of the spinal cord below shows an H-shaped mass of gray matter, which is divided into **'horns'**. Horns are mainly made up of neuron cell bodies.

- The cell bodies located in the two **posterior (dorsal) horns** primarily relay sensations and information.
- The two **anterior (ventral) horns** contain motor neurons and are thus involved in voluntary or reflex motor activity.

The horns are surrounded by white matter which consists of myelinated nerve fibers (axons) which are grouped into **'columns.'** All axons in one column perform the same general function.

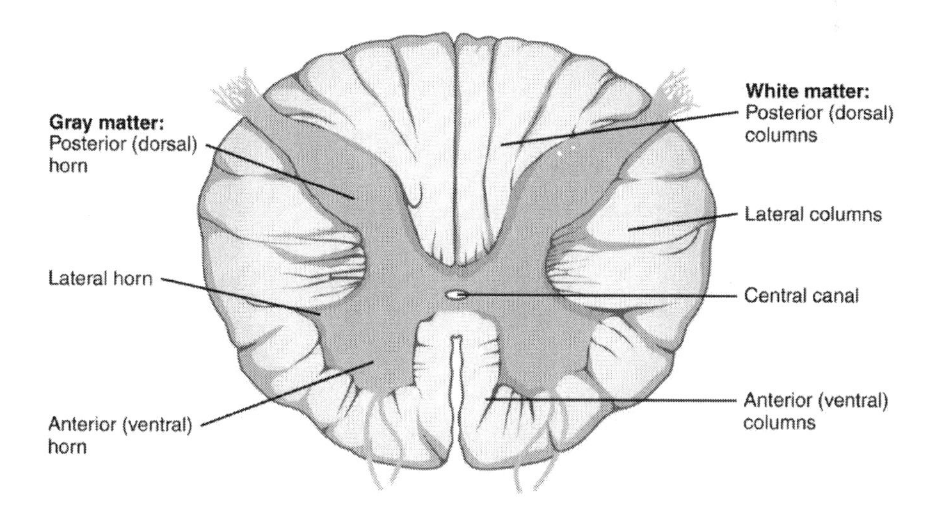

Sensory Pathways and Motor Pathways

Sensory impulses travel via **afferent neural pathways** to the sensory cortex located in the parietal lobe of the brain. It is in this lobe that the sensory impulses are interpreted. There are two afferent neural pathways:

- **The Ganglia:** Ganglia are relay stations composed of a number of nerve cell bodies located on the dorsal roots of spinal nerves. Sensations including pain, pressure, touch, and vibration enter the spinal cord via ganglia.
- **The Dorsal Horn:** Sensations including pain and temperature enter the spinal cord via the dorsal horn.

Motor impulses travel via **efferent neural pathways** from the brain to the muscle. Motor impulses are triggered in the motor cortex of the brain's frontal lobe and travel to lower motor neurons via upper motor neurons. These upper motor neurons form two main systems: the pyramidal system and the extrapyramidal system.

- **The Pyramidal System (Corticospinal Tract):** responsible for fine movements of skeletal muscles. Impulses in this system originate in the motor cortex, from which they travel through the internal capsule to the medulla, and down the spinal cord.
- **The Extrapyramidal System (Extracorticospinal Tract):** responsible for gross motor movements. Impulses in this system originate from the frontal lobes of the cerebrum, from which they travel through the pons, and down the spinal cord to the anterior horn. At the anterior horn, the impulses are relayed to the lower motor neurons, which then transmit these impulses to the muscles.

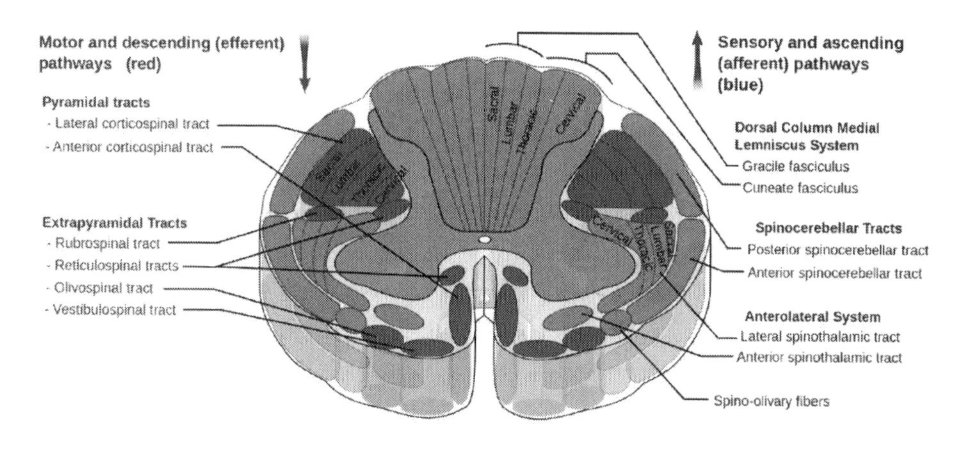

Structures That Protect the CNS

Meninges, comprising the dura mater, arachnoid membrane and pia mater, lines the vertebral canal and the skull. The meninges, together with cerebrospinal fluid, the skull and the vertebrae, protect the skull and the spinal cord from infection and shock.

- **Dura mater** is a tough and fibrous tissue which consists of two layers: the endosteal dura and the meningeal dura. The endosteal dura forms the periosteum of the bone (skull) and the meningeal dura is a thick membrane which protects the brain tissue. **Subdural spaces** are the spaces found between the dura mater and the arachnoid membrane.
- **Arachnoid membrane** is a thin and delicate membrane that lies between the dura mater and the pia mater. **Subarachnoid spaces** are found between the arachnoid membrane and the pia mater.
- **Pia mater** is a delicate and continuous layer of connective tissue, as well as the innermost membrane that envelops the brain and the spinal cord.

Cerebrospinal fluid (CSF) is a colorless body fluid which is found within the subarachnoid space, the ventricles, and the central canal of the spinal cord. CSF is composed of organic materials such as protein, glucose and electrolytes. CSF provides immunological protection to the brain and also serves as a 'buffer' for the brain's cortex, thereby protecting its tissue from blows.

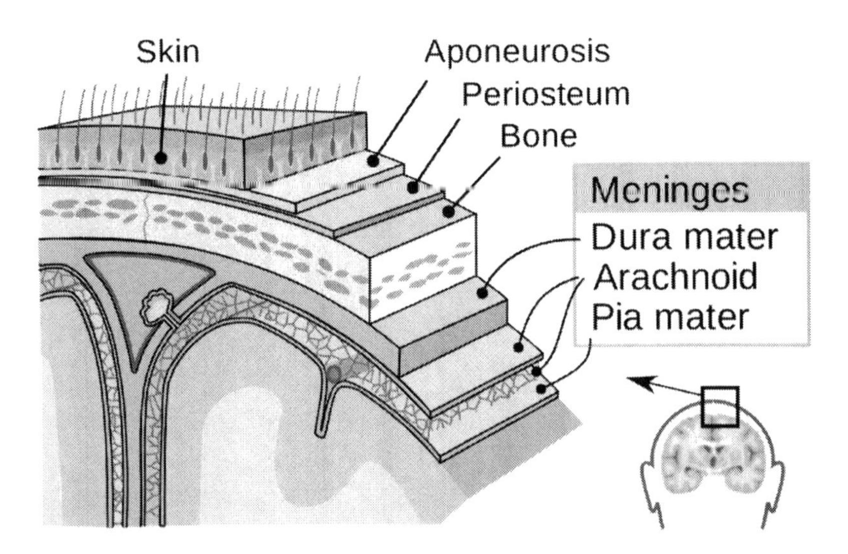

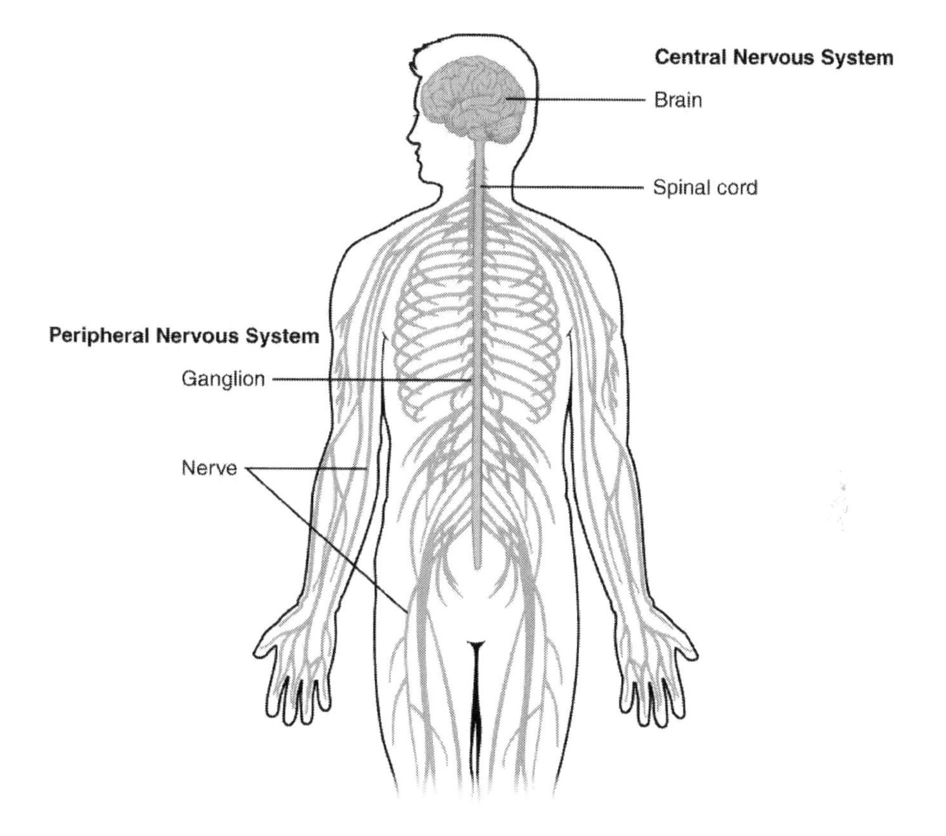
Central Nervous System
Brain
Spinal cord
Peripheral Nervous System
Ganglion
Nerve

The Peripheral Nervous System

The peripheral nervous system comprises the cranial nerves, spinal nerves, and autonomic nervous system (ANS).

Cranial Nerves:

There are twelve pairs of cranial nerves that transmit motor or sensory impulses between the brain or brain stem and the head and neck. 10 out of 12 cranial nerves exit from the brain stem. The two remaining pairs, the olfactory and optic nerves, exit from the forebrain. Each pair is dedicated to a particular function.

Paris of cranial nerves:

- **Olfactory (CN I)** - Sensory: Smell
- **Optic (CN II)** - Sensory: Vision
- **Trigeminal (CN V)** - Sensory: provides sensations to the skin of the face and Motor: controls the muscles of mastication (chewing)
- **Facial (CN VII)** - Sensory: taste receptors and Motor: facial muscle movement
- **Acoustic (CN VIII)** - Sensory: hearing and sense of balance
- **Glossopharyngeal (CN IX)** - Motor: swallowing and Sensory: oral sensation and taste
- **Vagus (CN X)** - Motor: controlling heart rate and food digestion and Sensory: sensations of the heart, lungs, bronchi, or GI tract
- **Spinal Accessory (CN XI)** - Motor: head rotation and shoulder movement
- **Hypoglossal (CN XII)** - Motor: tongue movement

Eye movement is coordinated by cranial nerves III, IV and VI:

- **Oculomotor (CN III)** - Motor: upper eyelid elevation, pupillary constriction, extraocular eye movement (superior, medial, and inferior lateral)
- **Trochlear (CN IV)** - Motor: extraocular eye movement (superior oblique)
- **Abducens (CN VI)** - Motor: extraocular eye movement (lateral)

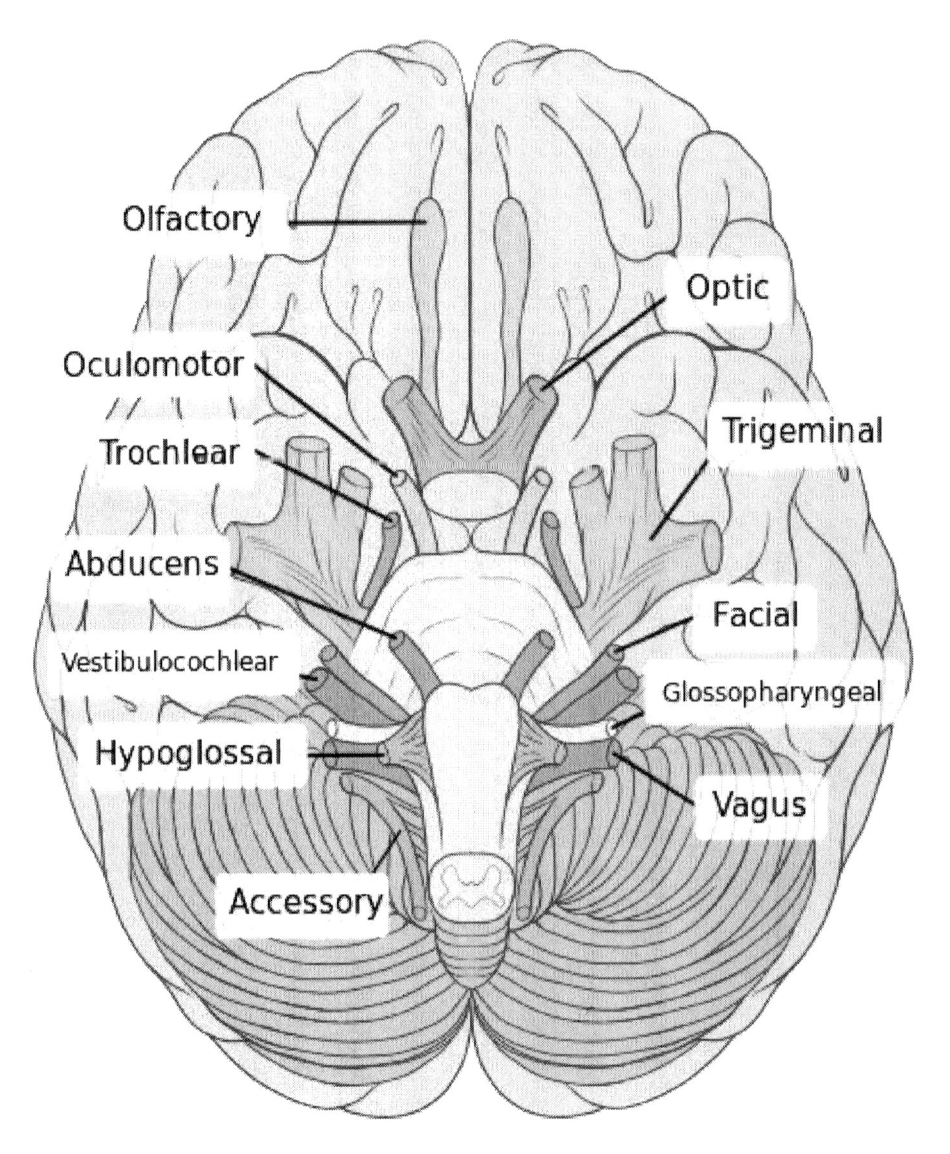

Spinal Nerves:

There are thirty-one pairs of nerves that emerge from the spinal cord. Each contains thousands of efferent (motor) fibers and afferent (sensory) fibers. These fibers carry messages to and from different body regions called dermatomes. The cervical nerves are designated C1 to C8. The thoracic nerves are classified as T1 to T12, the lumbar nerves are designated L1 to L5 and the sacral nerves are classified as S1 to S5.

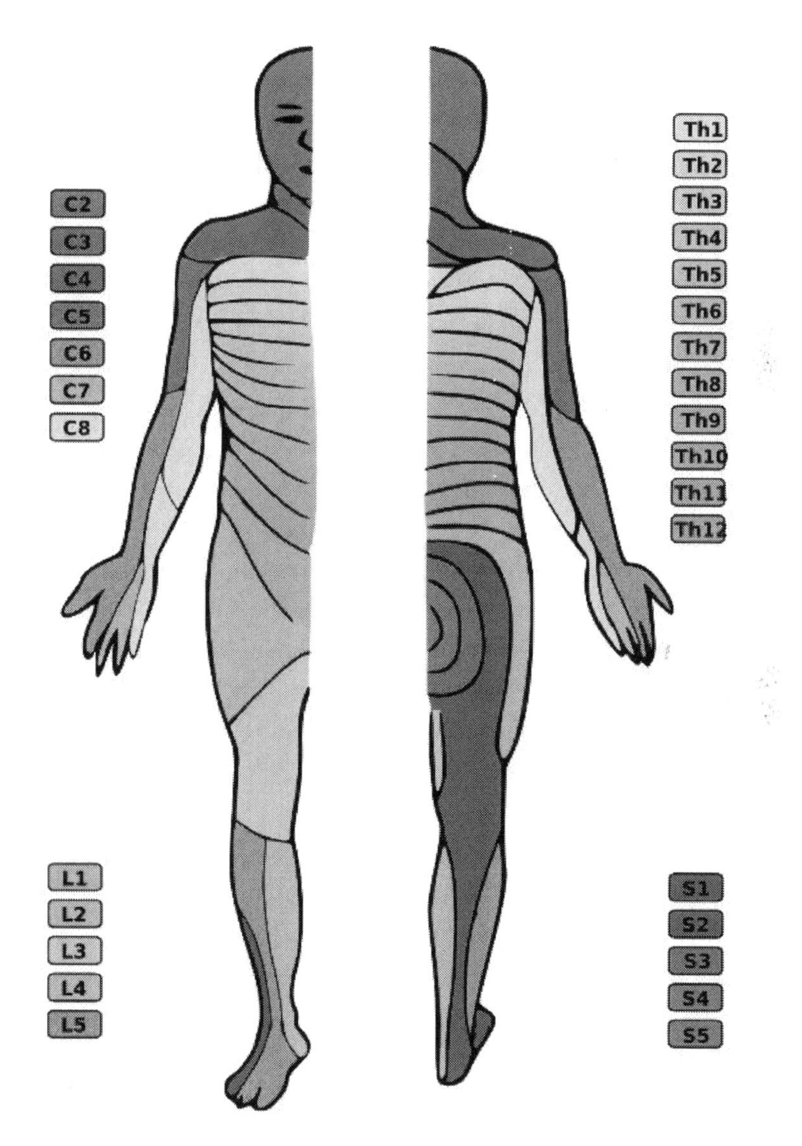

Autonomic Nervous System

The autonomic nervous system (ANS) supplies nerves to all internal organs. The ANS has to main subdivisions: the sympathetic **(thoracolumbar)** nervous system and the parasympathetic **(craniosacral)** nervous system. These two systems counterbalance each other, allowing the body systems to run smoothly.

Sympathetic Nervous System

Preganglionic neurons (sympathetic nerves) exit the spinal cord and then enter the ganglia. The ganglia transmit the impulse to postganglionic neurons which reach the organs and glands, triggering physiologic responses which include:

- Contraction of the sphincter
- High blood pressure
- Increased blood flow to skeletal muscles
- Increased heart rate and contractility
- Increased reparatory rate
- Increased sweat gland secretion
- Reduced pancreatic secretion
- Relaxation of the ciliary muscle and pupillary dilation
- Relaxation of the smooth muscle in the GI tract, bronchioles, and urinary tract
- Vasoconstriction

Parasympathetic Nervous System

Nerve fibers of the parasympathetic nervous system leave arise from the CNS. After leaving the CNS, the long preganglionic fiber of each parasympathetic nerve travels to a ganglion near a particular gland or organ. Short parasympathetic fibers create a more specific response in one gland or organ, which may include any of the following:

- Constriction of the bronchial smooth muscle
- Constriction of the pupil

- Increased bladder tone and urinary system sphincter relaxation
- Increased GI tract tone and peristalsis (with sphincter relaxation)
- Increased salivary, lacrimal, and pancreatic secretions
- Reduced heart rate and contractility
- Vasodilation of external genitalia

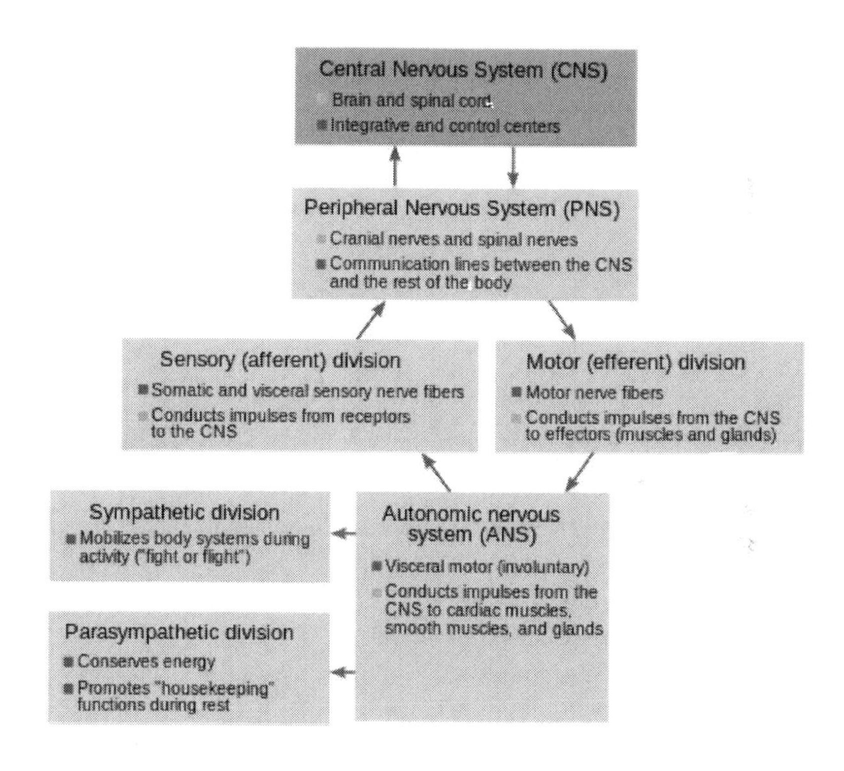

Special Sense Organs

Sense organs are distributed over the entire body and it is sensory stimulation which enables the body to interact with the environment. There are five sense organs - the eyes, ears, nose, tongue and skin - which contain general or special receptors that send messages to the brain. General receptors which are present in the skin, muscles, joints and visceral organs, are present throughout the body. Special receptors include **light receptors**

(also called photoreceptors) which are found in the eyes; **mechanoreceptors,** which are found in the ears; and **chemical receptors (chemoreceptors)**, which are found in the mouth and the nose.

The Eyes

The eye is the organ which gives us the sense of sight. Eye function is controlled by the extraocular and intraocular eye structures:

Extraocular Eye Structure

Extraocular eye structures include the extraocular muscles, eyelids, lacrimal apparatus, and conjunctivae. These structures protect and support the eyeball.

- **Extraocular Muscles:** The extraocular muscles control the motion of each eye and hold the eyeballs in place. Extraocular muscles have mutually antagonistic actions which means that as one muscles contracts, another (the antagonist muscle) relaxes.

- **Eyelids:** The eyelids, also referred to as the **palpebrae**, are the upper and lower folds of skin that cover the anterior portion of the eye (the exposed portion of the eyeball). The eyelids contain three different types of glands:
    - **The meibomian glands** are special sebaceous glands that secrete sebum which keeps the eyes lubricated.
    - **Glands of Zeis** are unilobal sebaceous glands which are connected to the follicle of the eyelashes and that service the eyelash.

- **Moll's glands** are ordinary sweat glands that are found on the margin of the eyelid and which cover the eyes completely when closed.

- **Lacrimal Apparatus:** The lacrimal apparatus comprises the lacrimal glands, lacrimal sac, punctum, and nasolacrimal duct. The lacrimal apparatus creates and absorbs tears that protect and lubricate the cornea and the conjunctivae. Furthermore, tears also contain an enzyme called lysozyme which protects the eye from bacterial invasion.

- **Conjunctivae:** The conjunctivae are mucous membranes that protect the eye from microbes that may otherwise enter the eye. They also help lubricate the eye and contribute to immune surveillance. The **bulbar (ocular) conjunctivae** cover the anterior portion of the sclera and the **palpebral conjunctivae** cover the inner surface of the eyelid.

Intraocular Eye Structure

Intraocular eye structures are directly concerned with vision. The intraocular eye structures are further divided into an anterior and a posterior segment.

The anterior segment includes the sclera, cornea, pupil, anterior chamber, aqueous humor, lens, ciliary body, and posterior chamber.

- **Sclera**: The sclera is the white outer layer of the eyeball which is continuous with the cornea. It maintains the size and form of the eyeball.
- **Cornea**: The cornea is the smooth and transparent layer which forms the front of the eye. The cornea is kept moist by tears and is highly sensitive to touch.

- **Iris**: The iris is a circular muscular ring that surrounds the pupil of the eye. Eye color is depended on the amount of pigment contained in the endothelial layer of the iris.
- **Pupil**: The pupil is the opening of the iris.
- **Anterior Chamber**: The anterior chamber is the cavity inside the eye that lies in front of the cornea's innermost surface and behind the iris.
- **Aqueous humor:** The anterior chamber is filled with aqueous humor, a clear and watery fluid.
- **Lens:** The lens is a transparent structure in the eye situated directly behind the iris at the pupillary opening. The lens, which is composed of transparent fibers called the lens capsule, helps refract and focus light on the retina.
- **Ciliary Body:** The ciliary body connects the iris to the choroid. It consists of three muscles which control the shape of the lens. Alongside the muscles of the iris, the ciliary body helps regulate light that is focused onto the retina.
- **Posterior Chamber:** The posterior chamber is the narrow space located behind the iris and in front of the lens. Like the anterior chamber, the posterior chamber is also filled with aqueous humor.

The posterior segment includes the vitreous humor, posterior sclera, choroid, and the retina.

- **Vitreous humor**: The vitreous humor is the thick transparent tissue of the eyeball that fills the space behind the lens.
- **Posterior sclera**: The posterior sclera is the white and fibrous layer that covers the posterior segment of the eyeball.
- **Choroid:** The choroid is a pigmented vascular layer which contains small veins and arteries. The choroid lies between the retina and the posterior sclera.
- **Retina:** The retina is the innermost layer of the eyeball. The retina is sensitive to light, receives visual stimuli, and triggers nerve impulses that are sent to the brain.

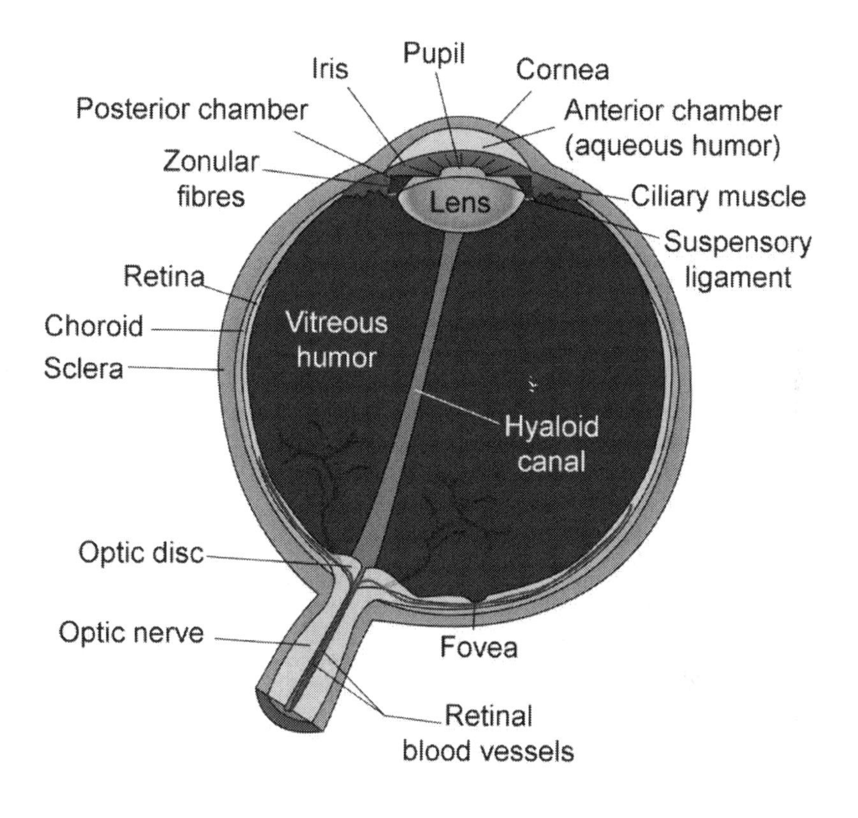

The retina contains the optic dish, physiologic cup, macula, rods and cones, fovea centralis, and four sets of retinal blood vessels.

- **Optic Disk:** The optic disk is the raised oval area in the retina which connects the retina to the optic nerve.
- **Physiologic Cup:** This is a funnel-shaped, light-colored depression within the optic disk. The central retinal blood vessels pass through the physiologic cup.
- **Rods and Cones:** The rods and cones are the visual receptors (the photoreceptor neurons) of the retina.

- **Macula:** This is the region surrounding the fovea of the eye. A light depression in the center of the macula is called fovea centralis. The macula is the area of highest visual acuity.
- **Fovea Centralis:** The fovea centralis is the small depression in the retina of the eye. Composed of closely packed cones, the fovea centralis is the main receptor for vision and color and constitutes the point at which visual acuity is greatest.

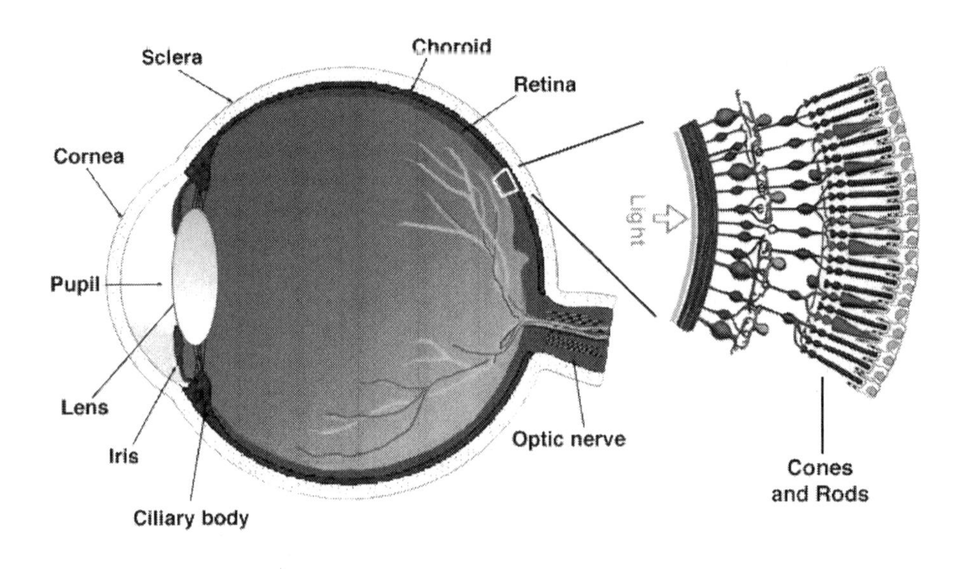

Basic Vision Pathways

The intraocular eye structures are involved in perceiving and forming images that are the sent to the brain for interpretation. To interpret images, the brain relies on structures which are found along the vision pathway. The vision pathway includes the retina, the optic nerves and the optic chasm.

Image formation occurs when eye structures (aqueous humor,

cornea, lens, and vitreous humor) refract light rays from an object, focusing it on the fovea centralis. Vision is generated by **photoreceptors (cones and rods)** in the **retina.** The visual information is transformed into impulses by the cones and rods and this impulse leaves the retina by way of the **optic nerves.** The optic nerves connect the eyeballs directly to the brain and send impulses to the brain for interpretation. Injury to one of the optic nerves can cause blindness in the respective eye.

The impulse travels from the optic nerves through the **optic chiasm** and into the optic section of the cerebral cortex. In the optic chiasm, which is a structure found in the forebrain, fibers from the left and right optic nerves cross over each other. Injury to the lesion in the optic chiasm can cause partial blindness. After the optic chiasm, information from the right visual field is sent to the left side of the brain and information from the left visual field is sent to the right side of the brain.

Some fibers from the temporal portions remain uncrossed. These uncrossed fibers form the **optic tract.** The optic tract, which is an extension of the optic nerve, is the pathway between the optic chiasm and the brain. There are two optic tracts: the left optic tract and the right optic tract.

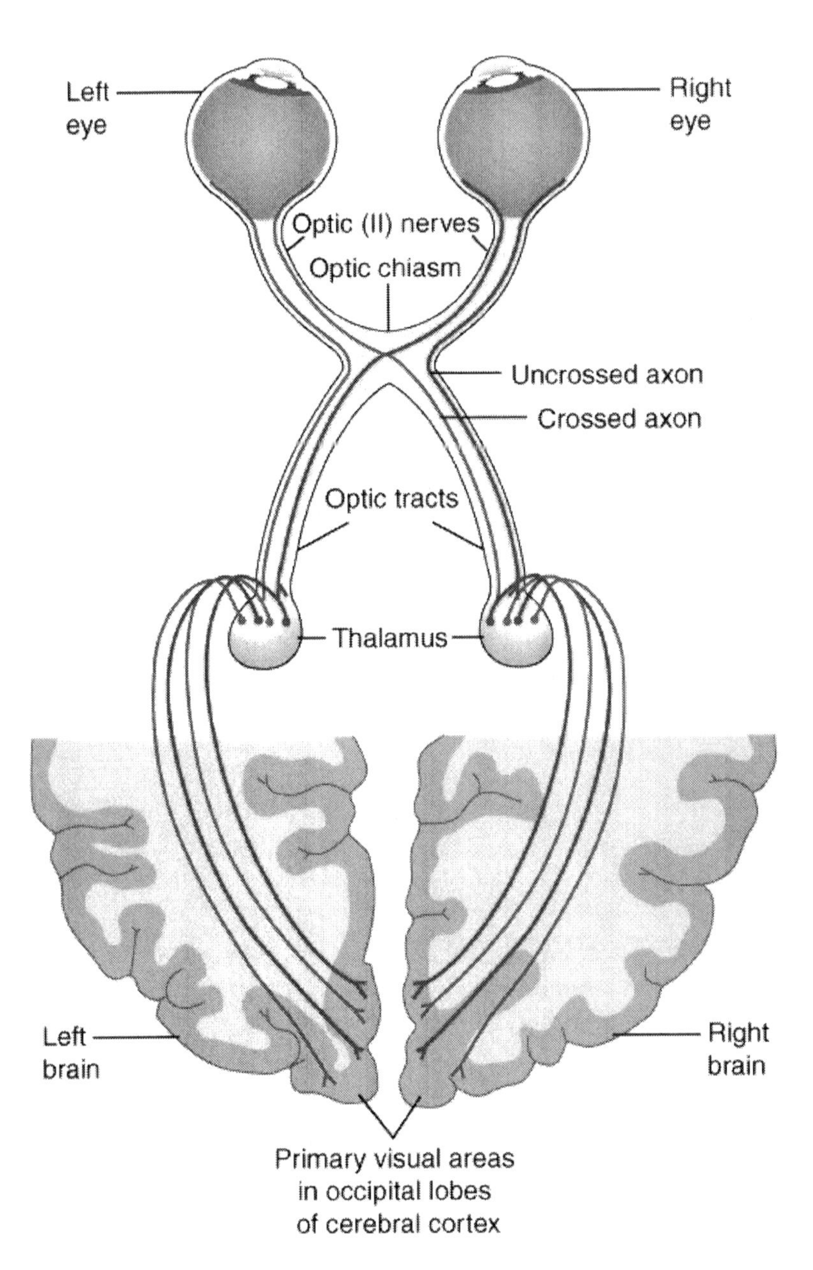
Left
eye
Right
eye
Optic (II) nerves
Optic chiasm
Uncrossed axon
Crossed axon
Optic tracts
Thalamus
Left
brain
Right
brain
Primary visual areas
in occipital lobes
of cerebral cortex

The Ears

The ears are the organs of hearing and balance. The ear is divided into three main sections: external, middle, and inner ear.

The External (Outer) Ear

The external ear consists of the pinna (auricle) and the external auditory canal.

- **The Pinna (Auricle):** The pinna is the external part of the ear in humans. It acts as a funnel to capture sound.
- **The External Auditory Canal:** The external auditory canal is the narrow chamber which connects the pinna to the tympanic membrane. It is also responsible for transmitting sound to the tympanic membrane and the eardrum.

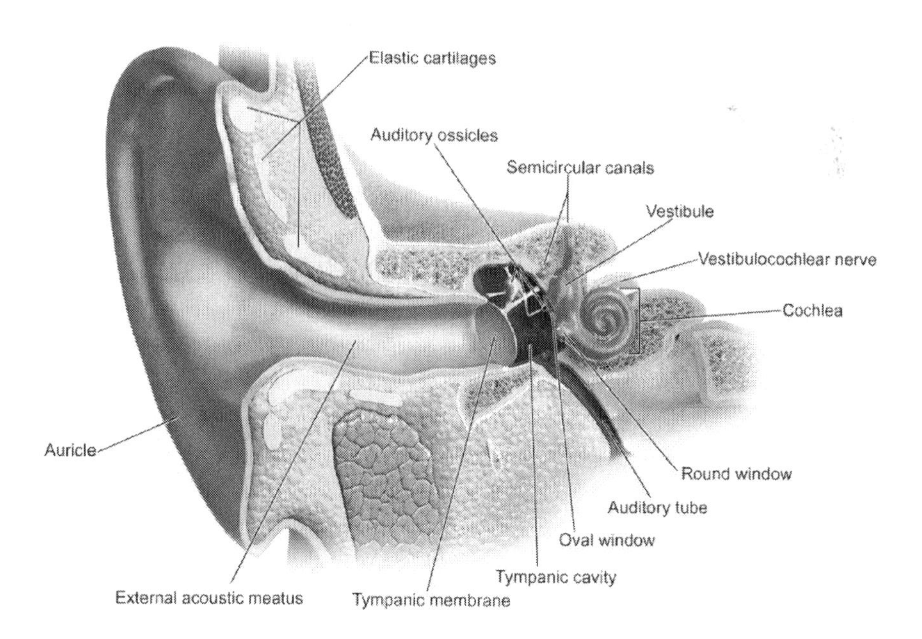

The Middle Ear

The middle ear, also called the tympanic cavity, is an air-filled space with structures. Structures in the middle ear include the tympanic membrane, Eustachian tube, oval window, round window, and small bones. The tympanic cavity is lined with mucosa, it is bound by the tympanic membrane on the distal end and by the oval window and round window on the medial end.

- **Tympanic membrane:** The tympanic membrane, which in humans forms the ear drum, is composed of layers of fibrous tissue, skin, and a mucous membrane. The tympanic membrane vibrates in responds to sound and transmits these sound vibrations to the inner ear.
- **Eustachian (Auditory) Tube:** The Eustachian tube, also called the auditory tube, extends from the middle ear to the nasopharynx (the upper part of the pharynx which is connected with the nasal cavity). The function of the auditory tube is to equalize the pressure on either side of the tympanic membrane. This prevents rupture while at the same time allowing the ear bones to vibrate.
- **(Auditory) Ossicles:** The ossicles are the small bones found in the middle ear which conduct vibrations from the tympanum to the oval window. The ossicles are the smallest bones found in the human body. The middle ear contains three ossicles: the **malleus (hammer)** transmits vibrations form the eardrum to the incus, the **incus (anvil)** transmits vibrations from the malleus to the stapes, and the **stapes (stirrup)** transmits vibrations from the incus to the inner ear. The function of the ossicles is to amplify sound vibrations.
- **Oval Window:** The oval window (*fenestra ovalis*) is the small hole opening in the wall between the middle and inner ear inside which the **stapes** (ossicle) is found.
- **Round Window:** The round window (*fenestra cochleae*) is also a small hole opening in the same wall. The round window is surrounded by the secondary tympanic membrane. The oval and round windows transmit vibrations to the inner ear.

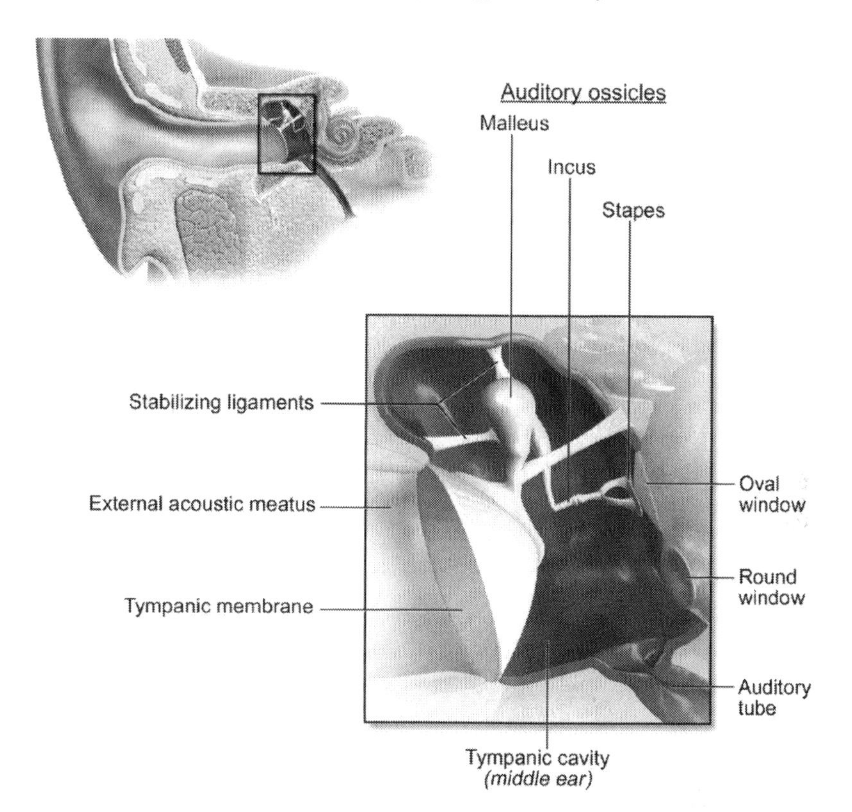

The Inner Ear

The inner ear is a complex structure found in the temporal bone. The **bony labyrinth** is the bony outer wall of the inner ear which consists of the vestibule, the cochlea, and the semicircular canals. Within the bony labyrinth is another structure called the membranous labyrinth. The **membranous labyrinth** is filled with a fluid called **endolymph.** The inner ear is filled with a fluid called **perilymph -** perilymph is the fluid between the bony labyrinth and the membranous labyrinth. Inside the inner ear, receptor nerve endings are stimulated by the vibrations.

- **Vestibule:** The vestibule, located behind the cochlea and before the semicircular canals, is the central part of the bony labyrinth. The vestibule is the entrance to the inner hear and houses two membrane sacs, the **saccule** and the

utricle, which are suspended in **perilymph.** The vestibular system plays a crucial role in maintaining balance.

- **Cochlea:** The cochlea is the bony spiral cavity of the inner ear which lies at the anterior part of the vestibule. Inside the cochlea is the **cochlear duct** which contains the **organ of Corti**, which is the organ responsible for hearing. The organ of Corti is the receptor organ for hearing which produces and transmits nerve impulses to the eight cranial nerve (CN VIII).
- **Semicircular canals:** The semicircular canals comprise three fluid-filled bony channels which lie at the posterior part of the vestibule. The semicircular canals are situated at right angles to each other with each individual canal orientated in one of three planes: **anterior, lateral** and **posterior.** The **semicircular ducts,** which are filled with endolymph and lined with microscopic hairs (cilia), traverse the canals and connects with the **utricle**. The utricle is a small otolith organ involved in sensing movement and gravity. At the end of each canals lies the **crista ampullaris.** The cristae ampullaris is a sensory organ which contains hair and support cells and which is responsible for sensing angular acceleration and deceleration.

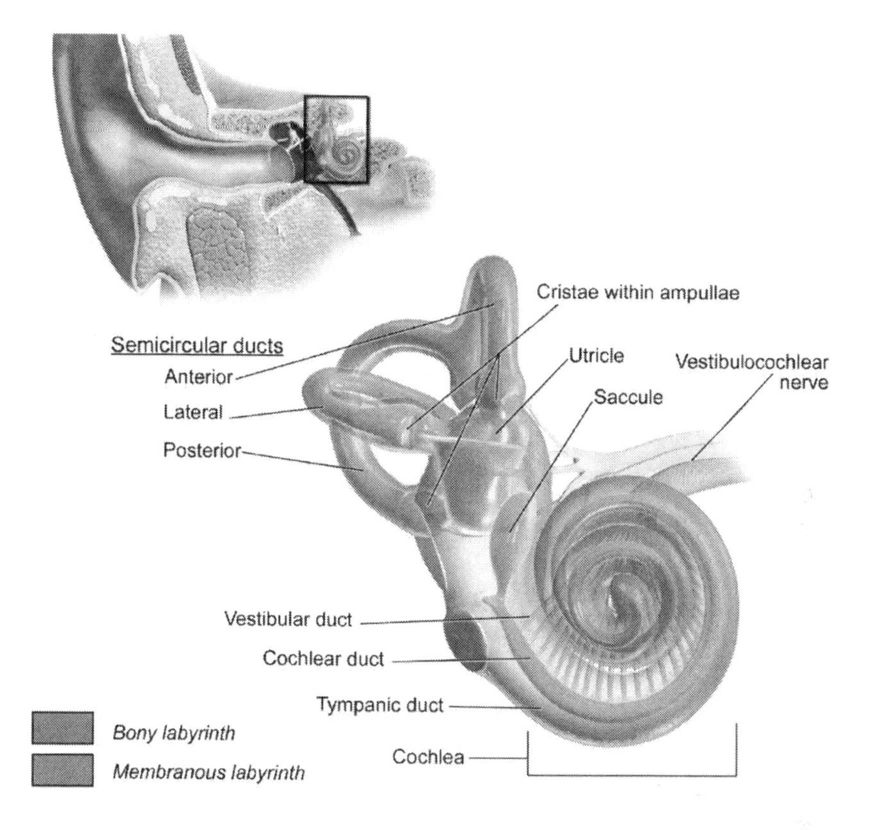

Basic Hearing Pathways

Sounds waves can travel through the ear by air conduction or bone conduction. In **air conduction**, hearing occurs through sound waves in the air near the ear. These waves travel to the inner ear through the external and middle ear. In **bone conduction**, hearing occurs through vibrations which travel through bone and into the inner ear.

In both pathways, vibrations received from the air or bone stimulate nerve impulses in the inner ear. The hearing (auditory) nerves carry these vibrations to the brain where nerve cells transmit these electrical signals to the auditory area of the cerebral cortex. Sound interpretation occurs in the cerebral cortex.

The hearing process, as illustrated by the below diagram, occurs as follows:

1. Sound enters through the outer ear,
2. The ossicles of the middle ear amplify the sound vibrations,
3. The organ of Corti organizes these sounds by frequency, and then produces and transmits nerve impulses,
4. Hearing nerves transmit the impulses form the cochlea to the brain stem,
5. The signals travel through the brain where they are being interpreted,
6. The impulse is received and recognized by the auditory cortex which processes the sound.

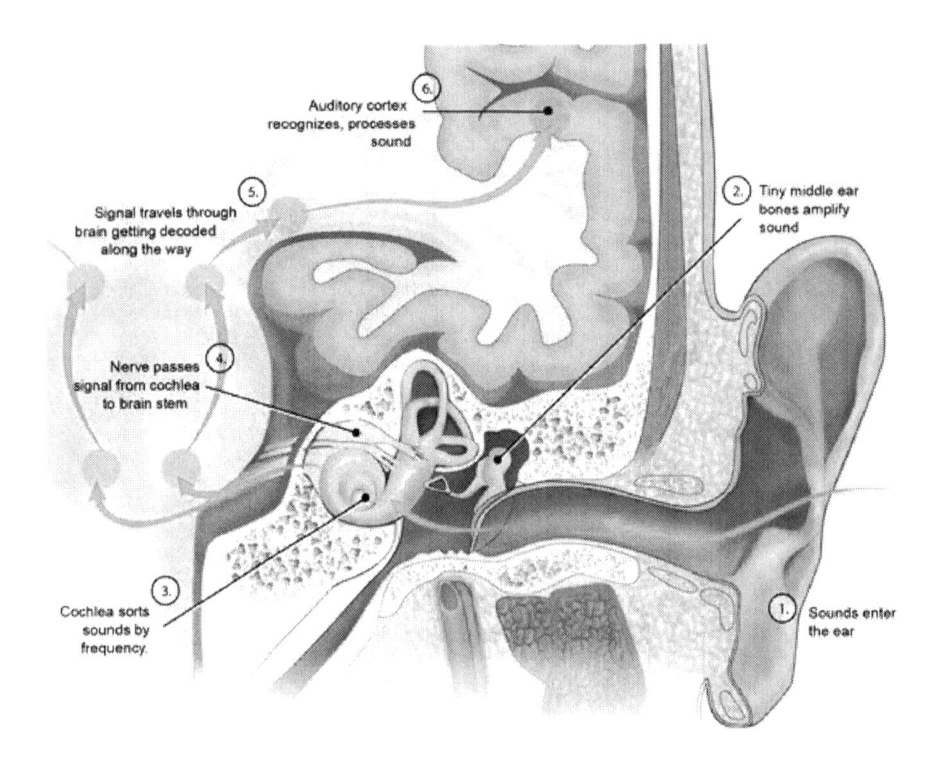

The Mouth

The majority of receptors for taste nerve fibers in the mouth are found on the tongue and on the roof of the mouth (CN VII and CN IX). These receptors, which are stimulated by chemicals, are also called taste buds. Taste buds provide for four senses of taste: sweet, sour, salty, and bitter. Other flavors are generated from a combination of stimulated taste buds and stimulated olfactory receptors.

The Nose

The nose is the sense organ for smell. Receptors for fibers of the **olfactory nerve (CN I)** are found in the mucosal epithelium which lines the upper portion of the nasal cavity. The olfactory nerves transmit impulses from the smell receptors to the brain. Smell receptors, also called **olfactory receptors**, are composed of highly sensitive hair cells which are stimulated by odors molecules.

The olfactory system includes the following structures:

1. **Olfactory Bulb:** The olfactory bulb is a brain structure involved in olfaction, that is, the sense of smell.
2. **Mitral Cells:** Mitral cells are neurons located in the olfactory bulb and which receive impulses from the axons of olfactory receptor neurons.
3. **Bone.**
4. **Epithelium:** Olfactory epithelium is epithelial issue which contains receptors for fibers of olfactory nerves.
5. **Glomerulus:** The glomerulus is a cluster of nerve endings located in the olfactory bulb.

6. **Olfactory Receptor Neurons:** Olfactory receptor neurons, also called olfactory sensory neurons, are responsible for odor detection.

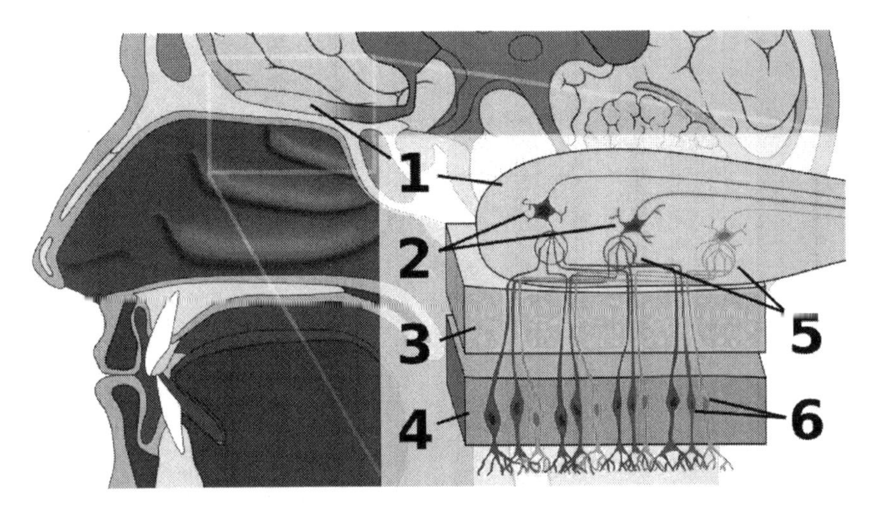

Section 7: The Integumentary System

The integumentary system is the largest body system, which includes the skin (integument) and its appendages (hair, nails, and particular glands).

The Functions of the Integumentary System

- Providing protection of the inner body structures.
- Sensory perception.
- The regulation of body temperature.
- The excretion of particular body fluids.

By migration and shedding, the skin supports the integrity of the body surface. By increasing the intensity of normal cell replacement mechanisms, the skin is able to repair surface wounds. The top layer of the skin is called the *epidermis,* which provides protection for the body against harmful chemicals and pathogen invasions.

Within the epidermis are specialized cells called Langerhans' cells. These increase the immune response from the body by assisting lymphocytes process antigens that enter the skin.

Melanocytes

Melanocytes are a type of skin cell, which provide protection for the skin by producing *melanin* (brown pigment). This helps with the filtering of ultraviolet (UV) light, which can stimulate the production of melanin.

Sensory Perception

Specific areas of the skin called *dermatomes* receive sensation from sensory nerve fibers, which are located in the nerve roots along the spine. A variety of sensations are transmitted by the nerve fibers to the skin, including temperature, touch, pain, pressure, and itching. Autonomic nerve fibers transport impulses to the smooth muscles in the skin's blood vessels, to muscles sitting around the hair roots, and to the sweat glands.

Temperature Regulation in the Body

The body's temperature (thermoregulation) is controlled by:

- Abundant nerves.
- Blood vessels.
- Eccrine glands.

All of these are located within the *dermis,* which is the deeper layer of the skin.

When the temperature of the body falls, as a result of the skin being exposed to cold, the blood vessels constrict, which decreases blood flow and conserves body heat. When the temperature of the body increases, the small arteries within the skin dilate, which increases blood flow and reduces body heat.

The Skins Excretion Function

Sweat is excreted by the sweat glands. This contains water, electrolytes, lactic acid, and urea. The skin excretes body wastes through over two million pores. Along with eliminating body wastes, the skin also stops fluids from the body escaping. This assists the body in preventing dehydration, which is caused by a loss of fluids. It also maintains the content and amount of sweat.

The skin also serves a function of preventing unwanted fluids in the exterior environment from entering the body.

The Layers of Skin

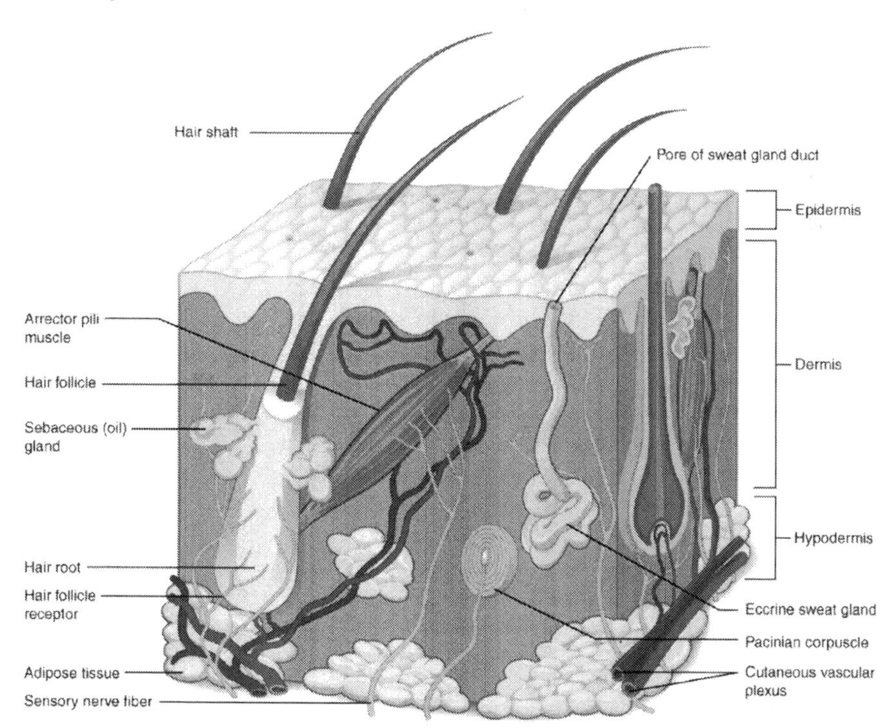

The *epidermis* and *dermis* are the two main layers of the ski, which lies above a layer called the *hypodermis*.

Epidermis

The epidermis is the external layer of the skin, which ranges in thickness from below 0.1mm on the eyelids, to over 1mm on the palms of the hand and soles of the feet. The epidermis allows light to pass through it, meaning it is translucent.

The epidermis is made up of stratified, avascular, and squamous epithelial tissue. It split up into five different layers:

- **Stratum Corneum (Horny Layer):** this is the outermost layer, which is made up of closely arranged cellular membrane and keratin layers.
- **Stratum Lucidum (Clear Layer):** this prevents water penetration or loss.
- **Stratum Granulosum (Granular Layer):** this forms keratin.
- **Stratum Spinosum (Spiny Layer):** also assists in forming keratin, and is contains ribonucleic acid.
- **The Stratum Basale (Basal Layer):** this is the innermost layer, which generates new cells that replace keratinized cells are have been shed or worn away.

Rete Pegs

There are no blood vessels in the epidermis, however the epidermis does contains fingerlike structures called *rete pegs.* Food, oxygen, and vitamins travel to the epidermis through the rete pegs, which are made up of a network of very small blood vessels that project down to the dermis layer.

Dermis

The second layer of the skin is called the dermis, which is an elastic system containing and supporting blood vessels, nerves, lymphatic vessels, and the epidermal appendages.

The majority of the dermis consists of an extracellular material called *matrix*, which contains the following:

- **Collagen:** this is a protein that is produced by the fibroblasts that provide strength and resilience to the dermis.
- **Elastic Fibers:** these make the skin flexible by binding the collagen.

The dermis is made up two layers:

- **Papillary Dermis:** this has fingerlike projections called papillae, which attach the dermis with the epidermis. These also help the gripping ability of fingers and toes.
- **Reticular Dermis:** this surrounds a layer of subcutaneous tissue. It consists of collagen fibers, providing the skin with strength, elasticity, and structure.

Subcutaneous Tissue

The subcutaneous tissue is the third layer of fat that lies below the dermis. It is made up of larger blood vessels and nerves, along with adipose cells filled with fat. The subcutaneous layer of fat lies on top of the muscles and bones, serving the function of energy storage, insulation, and shock absorption.

Epidermal Appendages

There are a number of epidermal appendages throughout the skin: the nails, hair, sweat glands, and sebaceous glands.

Hair

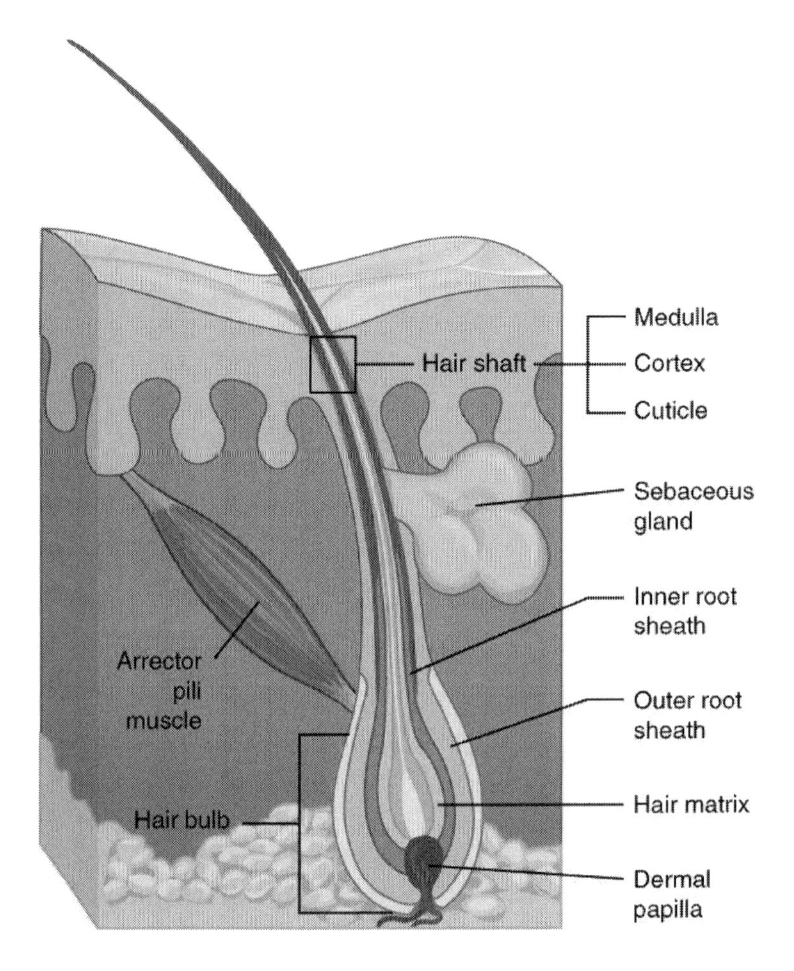

Hair is made up of keratin, and at each lower end of strand of hair is a *root*. Each root is indented by a *hair papilla*, which is an arrangement of blood vessels and cluster tissues. Each hair lies in a sheath lined with epithelium that is called a *hair follicle* that also have a supply of blood and nerves.

Nails

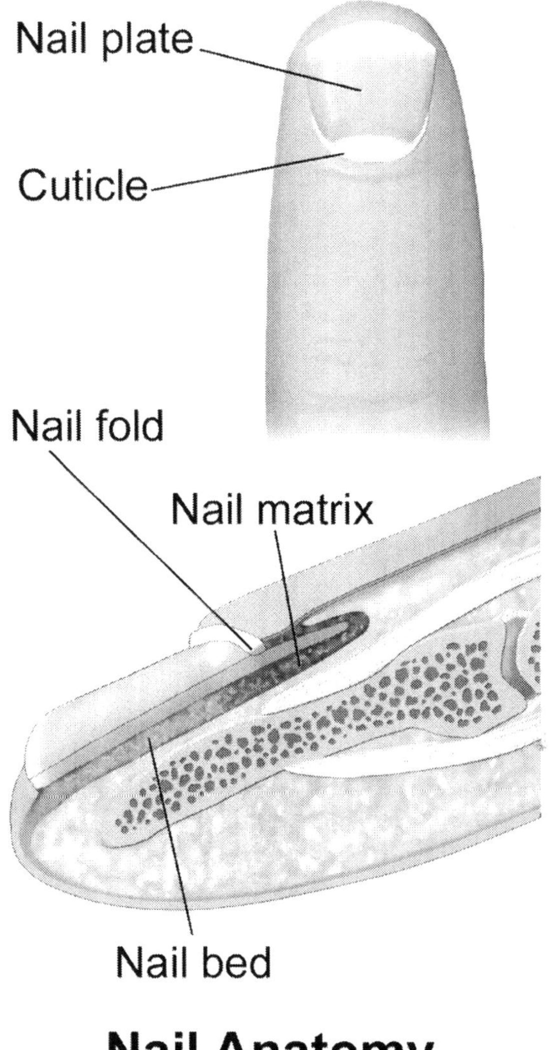

Nail Anatomy

The nails are located above the each finger and toe's distal surface. They are made up of a specialized type of keratin. The *nail plate* is positioned on the nail bed and surrounded by the cuticles (nail folds). The *nail matrix* extends approx. 0.5cm below the nail fold and forms the nail plate.

Sebaceous Glands

The sebaceous glands, which are part of the hair follicle, are on all parts of the skin, apart from the palms and soles. They produce *sebum,* which is a mixture of keratin, cellulose debris, and fat. When combined with sweat, sebum is an oily and moist acidic film that is partially antifungal and antibacterial. This exits through the hair follicle and provides protection for the surface of the skin.

Sweat Glands

The sweat glands are made up of two types of glands:

- **Eccrine Glands:** these are found throughout the body and they produce watery and odorless fluid. This fluid has a concentration of sodium that is equivalent to that of plasma. A duct from coiled secretory portion extends through the dermis and epidermis, which opens at the surface of the skin. Fluid from the eccrine glands is mostly secreted as a response to emotional stress.
- **Apocrine Glands:** these are found mainly in the underarm (axillary) and groin (anogenital) areas. In comparison to eccrine glands, their coiled secretory portion lies deeper in the dermis. The apocrine glands produce body odor as bacteria decomposes and they do not have a known biological function.

Section 8: The Endocrine System

The endocrine system comprises three main components: **glands** (specialized organs or clusters of cells), **hormones** (chemical substances which are secreted by glands), and **receptors** (protein molecules that bind to other molecules in order to stimulate physiologic changes).

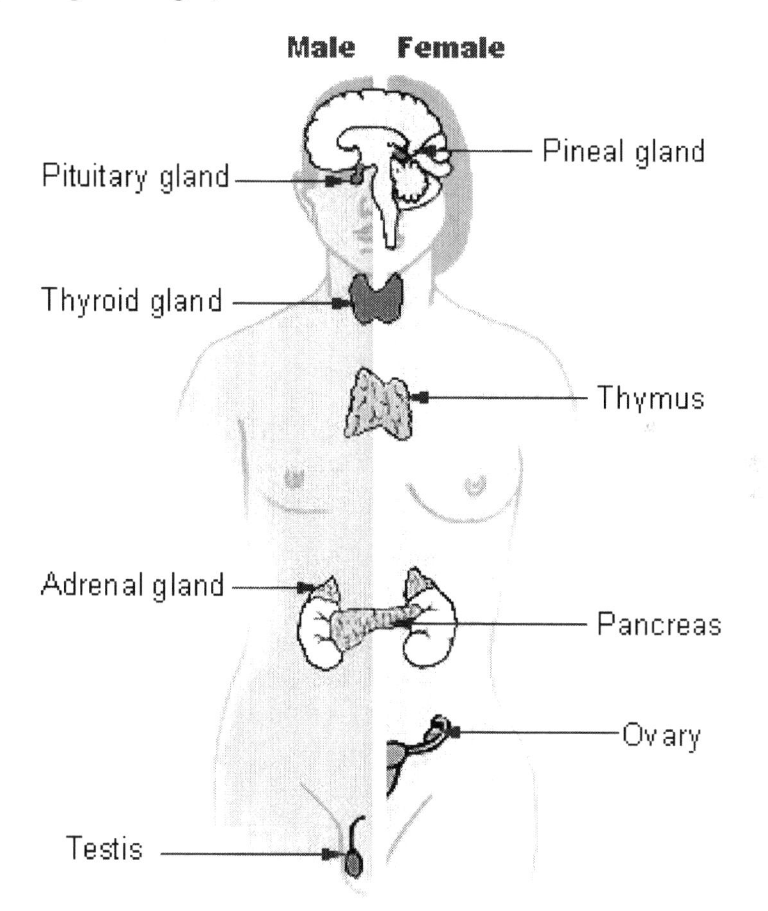

The glands found in the endocrine system include:

- Adrenal glands

- Gonads (testis or ovary)
- Pancreas
- Pineal gland
- Pituitary gland
- Thymus
- Thyroid and parathyroid glands

The Pituitary Gland

The pituitary gland, which is also called the hypophysis or master gland, lies in the **sella turcica**, a depression in the sphenoid bone that is found at the base of the skull. Messages are carried from the **hypothalamus** through the **infundibulum** (a funnel-shaped cavity) to the pituitary gland. The pea-sized pituitary gland comprises two regions: the **anterior pituitary** and the **posterior pituitary.**

The anterior pituitary, also called **adenophypophysis**, is the anterior lobe that together with the posterior lobe, makes up the pituitary gland. The anterior pituitary is the larger of the two and produces at least six hormones:

- Adrenocotrocotropic hormone (ACTH), sometimes called Corticotropin, which stimulates the production and release of cortisol from the adrenal gland.
- Follicle-stimulating hormone (FSH), which promotes the formation of sperm and ova.
- Growth hormone (GH), sometimes called Somatotropin (STH), which stimulates cell growth.
- Luteinizing hormone (LH), sometimes called Lutropin, which triggers ovulation and the development of the corpus luteum in females.
- Prolactin (PRL), which stimulates milk production after childbirth.

- Thyroid-stimulating hormone (TSH), sometimes called Thyrotropin, which stimulates and regulates the production of thyroid hormones.

The posterior pituitary is the back portion of the pituitary which secretes the following hormones:

- Antidiuretic hormone (ADH), sometimes called Vasopressin, which increases reabsorption of water by the kidney, and
- Oxytocin, which increases uterine contractions.

The Hypothalamus and the Pituitary Gland

Secretions by the anterior and posterior pituitary are controlled by the hypothalamus. As displayed in the diagram below, hypothalamic neurons manufacture inhibitory and stimulatory hormones. These hormones travel down to the anterior pituitary where they cause release or inhibition of pituitary hormones, including ACTH, TSH, GH, FSH, LH, and PRL.

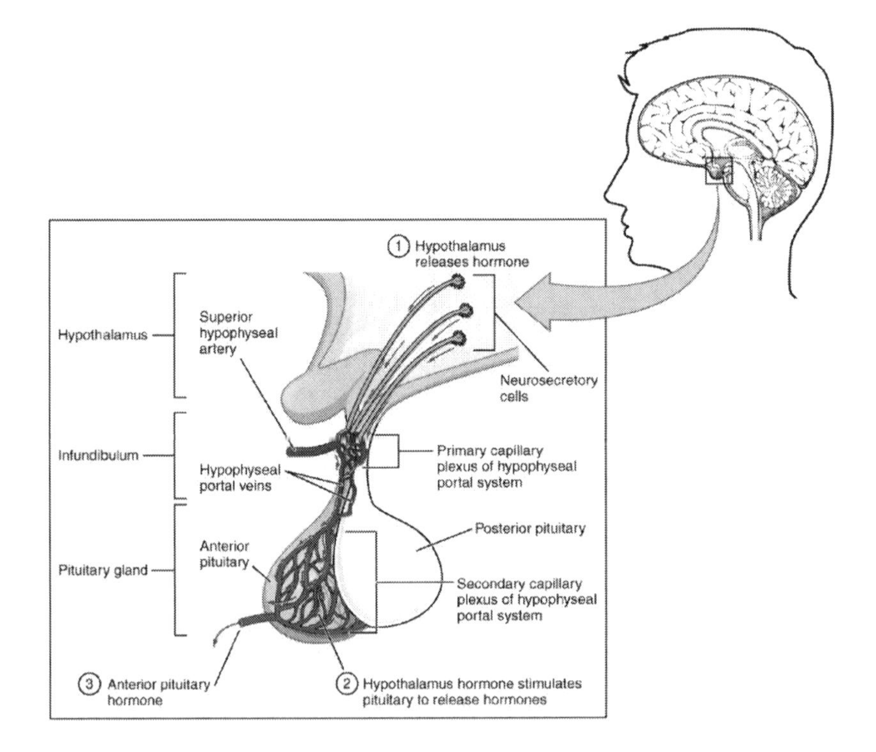

The Thyroid Gland

Thyroid hormones are essential for the functioning of every cell in the body. The thyroid gland is the gland which stores and produces hormones that help regulate blood pressure, body temperature, heart rate, and the rate of metabolism.

The thyroid is located directly below the larynx and has two lateral lobes which are joined by a narrow piece of tissue called the **isthmus.**

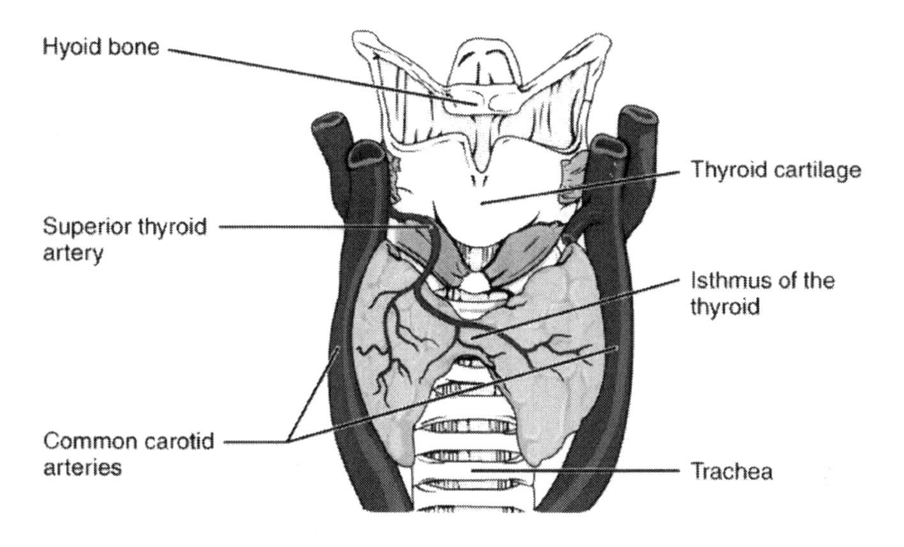

The two lobes function together to produce the following hormones:

- Triiodothyronine (T3),
- Thyroxine (T4), and
- Calcitonin.

Collectively, T3 hormones and T4 hormones are referred to as **thyroid hormones**. The thyroid hormone, which regulates the rate of metabolism by speeding up cellular respiration, is the principal metabolic hormone of the human body.

The Parathyroid Glands

The parathyroid glands, which are located on the posterior surface of the thyroid, are the smallest endocrine glands in the body. The parathyroid glands produce **parathyroid hormones** (PTH) which help regulate the body's calcium levels. PTH increases the movement of phosphate ions from the blood to the urine for excretion and also regulates the rate at which magnesium and calcium ions are lost via urine.

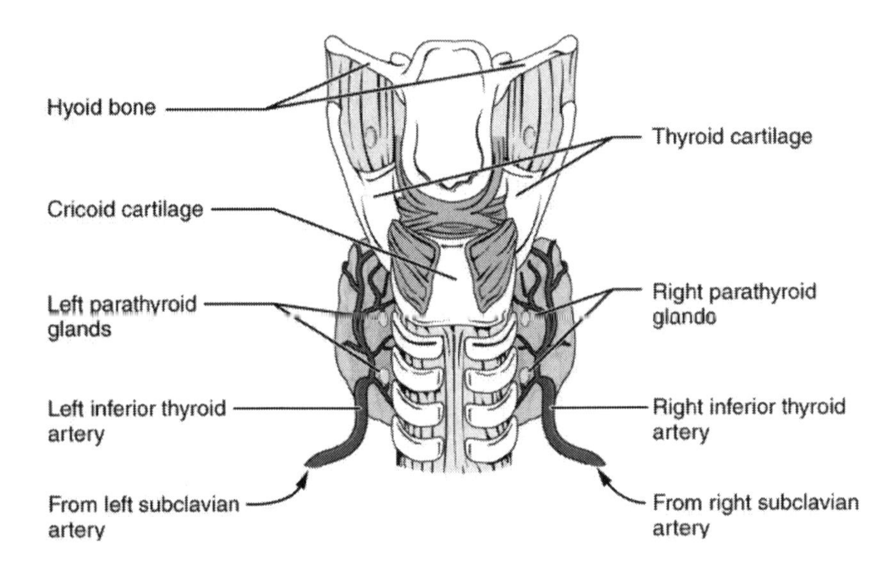

Adrenal Glands

The two adrenal glands of the body are located above the kidney. The adrenal glands are composed of two portions which function as separate endocrine glands: the adrenal cortex and the adrenal medulla.

Adrenal Cortex: The adrenal cortex is the outer portion of the adrenal gland. It contains three cell layers:

- The **zona glomerulosa** is the most superficial layer of the adrenal cortex, located directly below the renal capsule. It produces mineralcorticoids that help maintain fluid balance in the body by increasing sodium reabsorption.
- The **zona fasciculata** is the middle layer of the adrenal cortex. It produces glucocorticoids (which help regulate stress resistance and metabolism), cortisone,

corticosterone, as well as little amounts of androgen and estrogen (sex hormones).

- The **zona reticularis** is the innermost layer of the adrenal cortex which lies superficial to the adrenal medulla. It produces glucocorticoids and various sex hormones.

Adrenal Medulla: The adrenal medulla is the inner portion of the adrenal gland which is part of the sympathetic nervous system. The adrenal medulla secretes two hormones: epinephrine and norepinephrine (catecholamines), which are flight/fight hormones that play an important part in the autonomic nervous system.

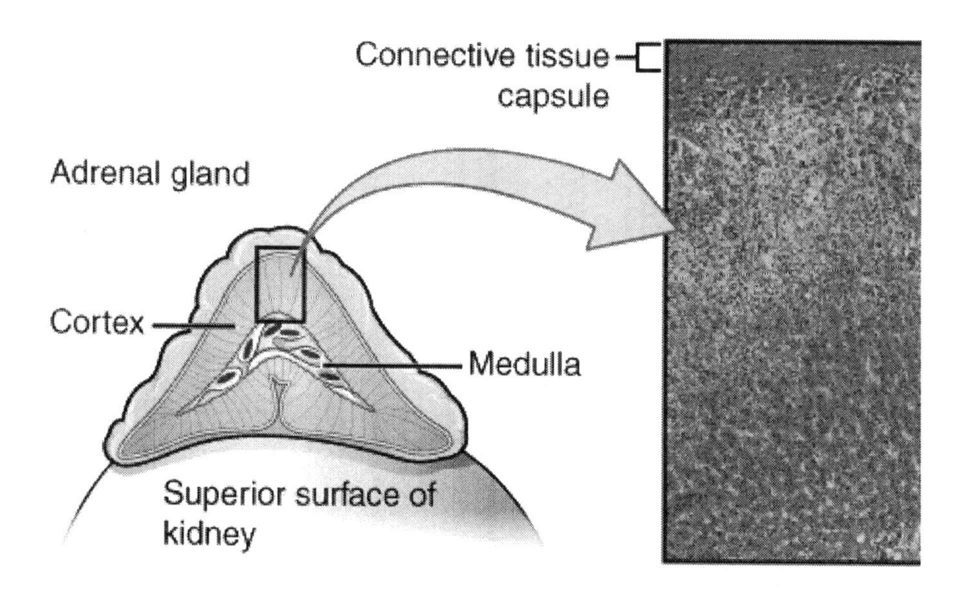

Tissue cross section continued below:

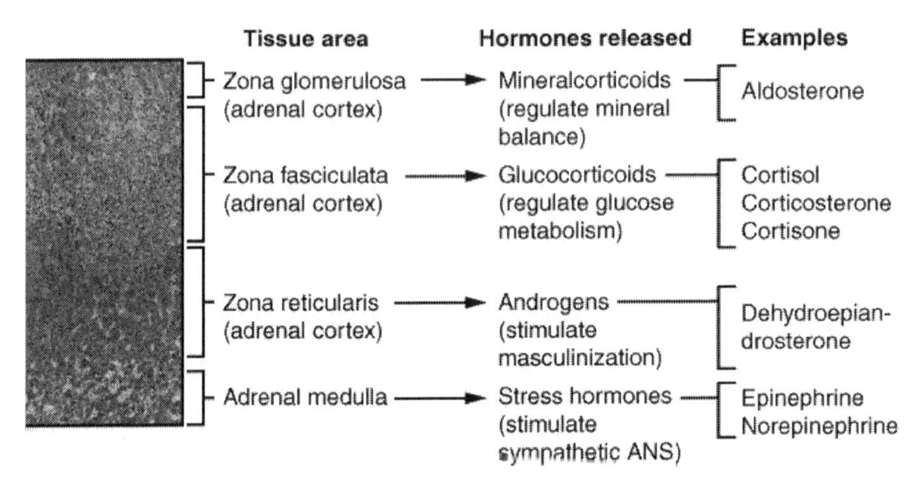

The Pancreas

The pancreas is a large gland located behind the stomach and which extends to the spleen. The pancreas performs exocrine functions (secreting digestive enzymes into the small intestine or gut) and endocrine functions (secreting hormones into the bloodstream). Pancreatic exocrine functions are controlled by **acinar cells**, which make up most of the pancreas.

The endocrine cells of the pancreas are called **islets of Langerhans.** Islet cells are found in in pancreatic tissue alongside acinar cells and exocrine cells. Islet cells contain alpha, beta, and delta cells which produce important hormones:

- **Alpha cells** produce glucagon which helps raise blood glucose levels by promoting the breakdown of glycogen to glucose in the liver.
- **Beta cells** produce insulin which helps regulate and lower blood glucose levels by stimulating the conversion of glucose to glycogen.
- **Delta cells** produce somatostatin which inhibits the release of certain hormones including corticotropin and GH.

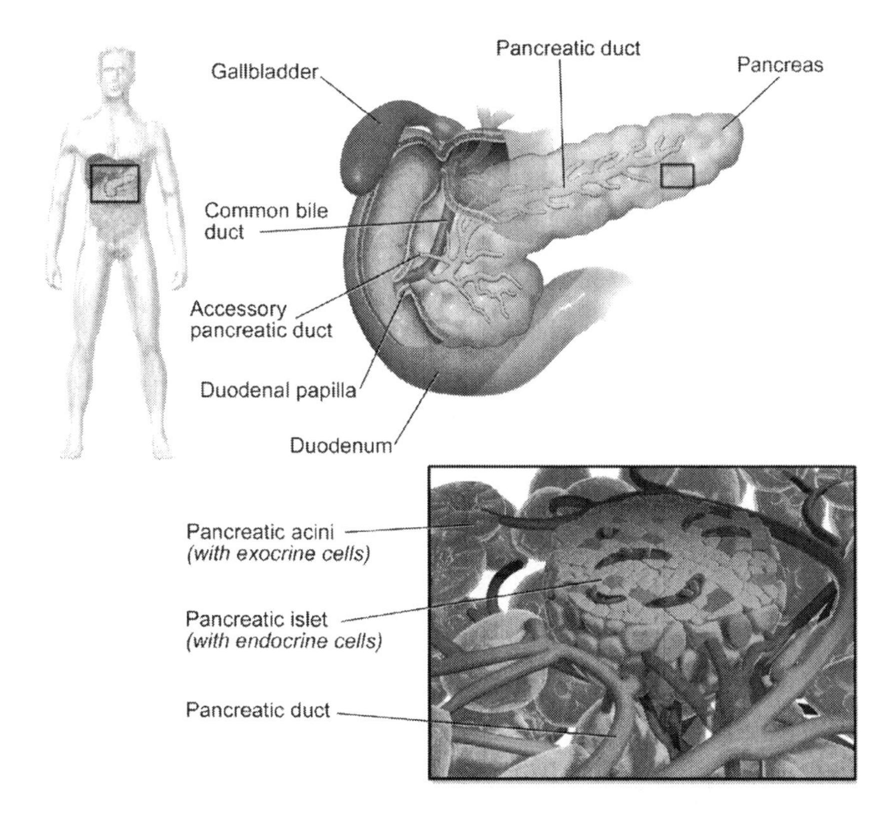

The Thymus

The thymus, which is located below the sternum, contains lymphatic tissue and secretes several hormones. The thymus is vital in the development of T cells which play a major role in the immune system. Besides T cells, the thymus also produces thymosin and thymopoietin.

The Pineal Gland

The pineal gland is a small gland located behind the third

ventricle of the brain. It produces the hormone melatonin which is involved in regulating sleeping and waking cycles, body temperature, circadian rhythms, reproduction, and cardiovascular function.

The Gonads

The gonads are the organs that produce gametes, namely the testes and the ovaries.

- **The ovaries** are oval-shaped glands that produce ova (eggs), estrogen and progesterone. These hormones have several functions which include regulating the menstrual cycle, maintaining the uterus for pregnancy, and promoting the development of female secondary sex characteristics. The ovaries also help to prepare mammary glands for lactation.

- **The testes** produce spermatozoa (male reproductive cells) and testosterone (male sex hormones). Testosterone maintains male sex characteristics and stimulates sex drive.

Hormones

Hormones are regulatory substances, small chemical messengers, that travel through the bloodstream triggering and regulating tissue or organ activity. Hormones are classified according to their molecular structure as **amines, polypeptides,** or **steroids.**

Amines: Amines are organic compounds derived from tyrosine, which is an amino acid that is a constituent of most proteins. The following hormones are amines:

- T3 and T4 (thyroid hormones), and
- Dopamine, epinephrine, and norepinephrine (catecholamines).

Polypeptides: Polypeptides are protein compounds composed of amino acids connected by peptide bonds. The following hormones are polypeptides:

- GH, TSH, FSH, LH, and ACTH (anterior pituitary hormones),
- ADH and oxytocin (posterior pituitary hormones),
- PTH (parathyroid hormone), and
- Insulin and glucagon (pancreatic hormones).

Steroids: Steroids, which are derived from cholesterol, contain four rings of carbon atoms. The following hormones are steroids:

- Aldosterone and cortisol (adrenocortical hormones which are secreted by the adrenal cortex), and
- Estrogen, progesterone, and testosterone (sex hormones secreted by the gonads).

Hormone Transport

Hormones are transported throughout the body by the bloodstream. When they reach their target organ, **polypeptides** and some **amines** will bind to the membrane on the receptor site. Some smaller **steroids** and thyroid hormones which are lipid-soluble will diffuse through the cell membrane and bind to intracellular receptors. Once bound to a receptor, each hormone will produce specific physiologic changes. A hormone will only act on a cell that has receptors which are specific to the hormone.

Hormone Release

There are four basic mechanisms which allow the body to control the release of hormones: the pituitary-target gland axis, the hypothalamic-pituitary-target gland axis, chemical regulation, and nervous system regulation.

Pituitary-Target Gland Axis: The pituitary gland secretes **trophic hormones** which are releasing or inhibiting hormones. These releasing hormones and inhibiting hormones target other endocrine glands where they regulate hormone release. These hormones include:

- **Corticotropin** - regulates release of adrenocortical hormones.
- **TSH** - regulates release of T3 and T4.
- **LH** - regulates release of gonadal hormones.

The pituitary gland receives feedback about the target glands through monitoring the levels of hormones produced by these glands. Depending on the feedback received, the pituitary gland responds by:

- Increasing the trophic hormones which stimulate the target glands to increase their production of the target hormone.

- Reducing the trophic hormones which in turn decreases target gland stimulation and thus the levels of the target hormone.

Hypothalamic-Pituitary-Target Gland Axis: The hypothalamus also produces trophic hormones which target the anterior pituitary gland. It therefore regulates anterior pituitary hormones which in turn regulate target gland hormones.

Chemical Regulation: Specific chemicals trigger hormone secretion. Within the body, hormone secretion occurs in response to chemical signals which are received from several regulatory systems.

The Nervous System: The nervous system also regulates hormone secretion. Nervous system stimuli for example, such as pain, can trigger ADH levels. And stress for example, which results in a sympathetic simulation, can trigger corticotropin release.

Section 9: The Cardiovascular System

The cardiovascular system is made up of the heart, blood vessels, and the lymphatic system. The system serves the responsibility of providing oxygen and nutrients to the cells of the body. On top of this, it eliminates metabolic waste and transports hormones within the body.

The Heart

The heart is made up of two distinct pumps. The right side of the heart pumps oxygenated blood to the lungs, and the left side of the heart pumps and provides blood to the rest of the body.

The heart is located below the sternum in the cavity between the lungs (the mediastinum), between the 2nd and 6th ribs.

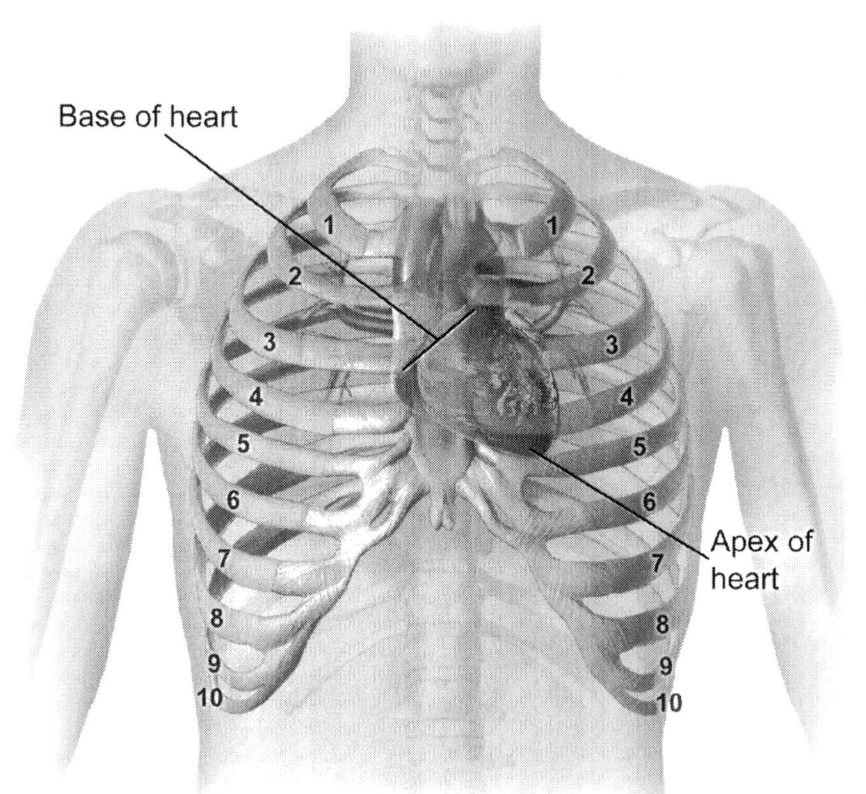

Heart Position Relative to the Rib Cage

The heart is positioned obliquely in most people - its right side below and nearly in front of the left side. Because of the angle it sits at, the top (base) of the heart is at its top right, and the pointed end (apex) is as its bottom left. The point of maximal impulse is at the apex. The heart sounds loudest at this point.

The Structure of the Heart

A membrane called the *pericardium* surrounds the heart. The heart has a wall that consists of the *myocardium, epicardium,* and the *endocardium.* Inside the heart there are four chambers: two ventricles and two atria, and four valves: two semilunar valves and two atrioventricular valves).

The Pericardium

The pericardium is a double-walled fibroserous sac that encloses the heart and the roots of the vessels that bring blood to and from the heart. It is made up of both fibrous and serous *pericardium.* The fibrous pericardium, which is made up of tough and white fibrous tissue, covers the heart thus providing it with protection. The serous pericardium is the smooth inner portion that has two layers: the parietal layer and the visceral layer.

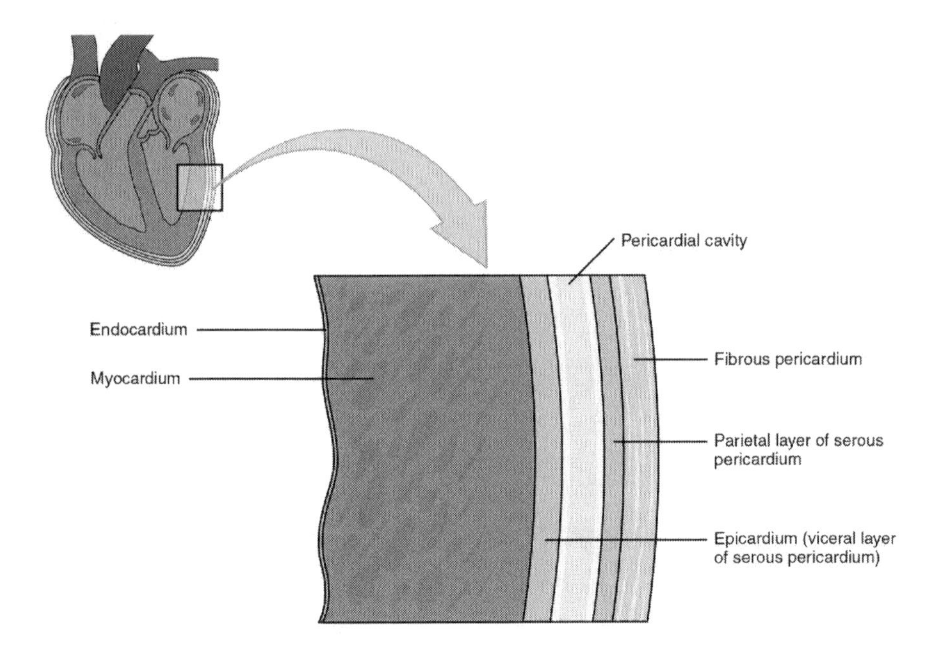

Sitting between the fibrous pericardium and serous pericardium is the pericardial cavity. Contained in the pericardial cavity is pericardial fluid, which plays the role of lubricating the surfaces and allowing the heart to move freely throughout contraction.

The Wall of the Heart

The wall of the heart is made up of three separate layers:

- **The Epicardium:** this is the outer layer, which consists of squamous epithelial cells that overlay the connective tissue.
- **The Myocardium:** this is the middle layer, which makes up the majority of the wall of the heart. It contains striated muscle fibers that can produce heart contractions.
- **The Endocardium:** this is the inner layer, which is made up of endothelial tissue, blood vessels, and smooth muscle.

Inside the Heart

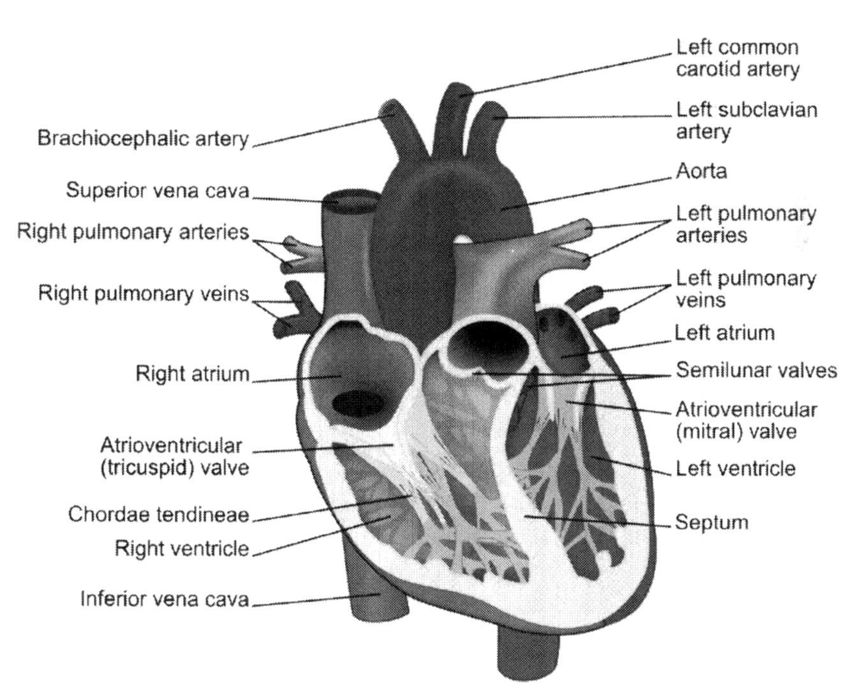

Inside the heart there are four chambers: two ventricles and two atria, and four valves: two semilunar valves and two atrioventricular valves). There is a system of blood vessels that carry blood in and out of the heart.

The Chambers

The interatrial septum separates the atria. The atria receive the blood that is returning to the heart, and they supply blood to the ventricles.

The Right Atrium: this receives blood from both the *superior and inferior and inferior venae cavae.*

The Left Atrium: this is smaller than the right atrium, but has walls that are thicker. It receives blood from the both of the pulmonary veins.

The Ventricles

The interventricular septum separates the right and left ventricles, which make up the lower chambers of the heart. They are made up of developed musculature and receive blood from the atria.

The Right Ventricle: responsible for pumping blood to the lungs.

The Left Ventricle: larger in size than the right ventricle, it pumps blood throughout all other vessels in the body.

The Valves

The valves let blood flow forwards through the heart and stop the backwards flow of blood. The valves open and close as a result of changes in pressure that are caused by ventricular contraction and the ejection of blood. The atria are separated from the ventricles by the two atrioventricular valves.

The Right Atrioventricular Valve (Tricuspid Valve): stops backwards blood flow from the right ventricle to the right atrium.

The Left Atrioventricular Valve (Mitral Valve): stops backwards blood flow from the left ventricle into the right atrium.

The Two Semilunar Valves: the pulmonic valve prevents backwards blood flow from the pulmonary artery to the right ventricle. The aortic valve prevents backwards blood flow from the aorta to the left ventricle.

Conduction System

The conduction system in the heart causes contractions that transport blood throughout the body. Fibers produce electrical impulses in the heart's cells that cause contractions of the heart. Below is a diagram of the cardiac conduction system that shows the different elements of the system.

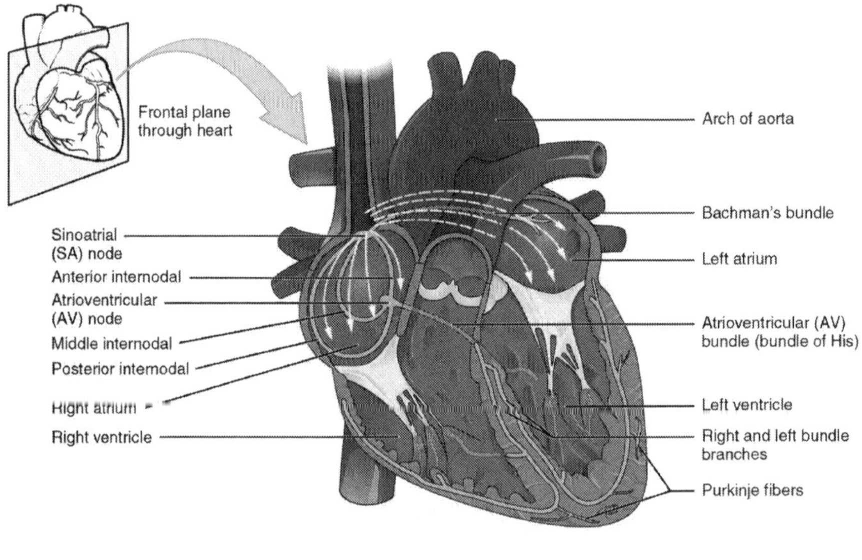

Anterior view of frontal section

The heart's conduction system contains pacemaker cells that have the following three features:

- **Automaticity:** generating automatic electrical impulses.
- **Conductivity:** passing impulses to the next cell.
- **Contractility:** shortening the heart's fibers when it is receiving impulses.

The Sinoatrial (SA) Node

This is the normal pacemaker of the heart that generates impulses between 60 to 100 times per minute. The SA node creates an impulse that spreads through both the right and left atria. The result of this is *atrial contraction.*

The Atrioventricular (AV) Node

The AV node is located in the lower portion of the right atrium's septal wall, and slows down the impulse conduction between the atria and the ventricles. This is a *resistor node* that allows for a time delay, which allows blood to be transferred from the atria to the ventricles.

Impulses travel from the AV node to the *bundle of His,* which are modified muscle fibers. Following this, they then branch of to the right and left bundle branches, and lastly travel to the *Purkinje fibers.*

There are two safety mechanisms that the conduction system has. Firstly, if there is a failure to fire an impulse from the SA, the AV node will generate an impulse between 40 and 60 times per minute. Secondly, if both the SA and AV node fail to fire, the ventricles are able to generate an impulse that is between 20 and 40 times per minute.

Cardiac Cycle

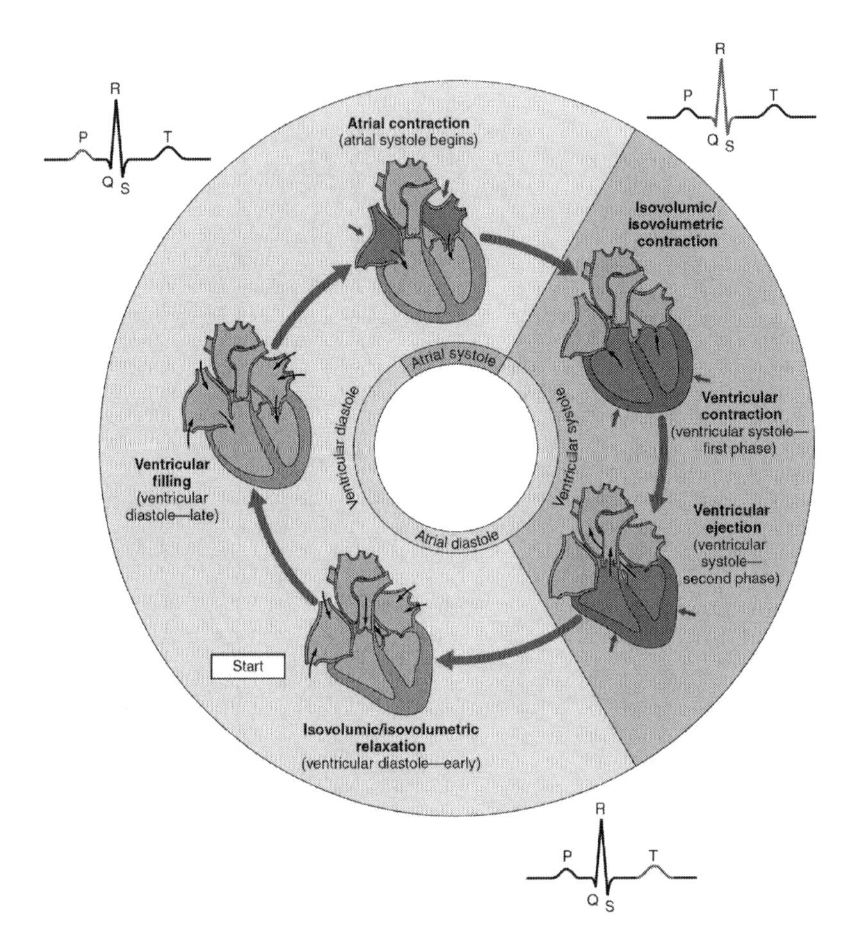

The period between the start of one heartbeat to the start of the next heartbeat is called the *cardiac cycle.* In order to provide the correct cardiac output, electrical and mechanical events must take place in the correct sequence and to the correct degree. The cardiac cycle is made up of two phases:

Systole:

In the beginning of the cardiac cycle (the systole) the ventricles contract. The increase in blood pressure in the ventricles makes the atrioventricular valves close, and the semilunar valves open. The ventricular blood pressure rises as the ventricles contract, and this happens until the pressure is greater than the pulmonary artery and aorta pressure. Following this, there is the opening of the semilunar valves, and blood is ejected by the ventricles into the pulmonary artery and aorta.

Diastole:

Once the ventricles are empty and become relaxed, the ventricle pressure decreases below the pressure in the aorta and the pulmonary artery. This is the beginning of the *diastole,* in which the semilunar valves close to prevent the backwards flowing of blood into the ventricles. There is the opening of the mitral and tricuspid valves, and this allows blood to flow from the atria into the ventricles. The atria contract and deliver the remaining blood to the ventricles when the ventricles become full. The heart enters systole and the cardiac cycle is restarted.

Cardiac Output

The cardiac output is the volume of blood that the heart pumps over a minute. The *stroke volume* is the volume of blood that is ejected from each heartbeat, and the cardiac output is calculated by multiplying the heart rate by the stroke volume. The stroke volume is dependent on preload, afterload, and contractility.

Preload: this is when the muscle fibers in the ventricles are stretched, which is a consequence from ventricle blood volume at the end of the diastole. The more the ventricles stretch, the more vigorously they will contract during systole.

Contractility: this is the myocardium's ability to contract normally. The preload influences this ability.

Afterload: this is the pressure that the ventricular needs to produce to deal with the elevated pressure in the aorta, in order to get blood out of the heart.

Blood Flow

Whilst travelling through the vascular system, blood makes its way through five different types of blood vessel: *arteries, arterioles, capillaries, venules,* and *veins*. Each vessel differs in relation to the function it serves in the cardiovascular system, and the pressure that is made by the blood volume at a number of points within the system.

Arteries have thick and muscular walls that support blood flowing quickly with high pressures. Arterioles have walls that are thinner than those of arteries – they control blood flow to the *capillaries* by constricting and dilating. The walls of the capillaries are made up of one layer of endothelial cells.

Blood from the capillaries is gathered by the venules. The walls of the venules are thinner than the arterioles walls. *Veins* have walls that are thinner than those of the arteries, however they have a diameter that is larger.

Below is an illustration of the major arteries and veins in the body.

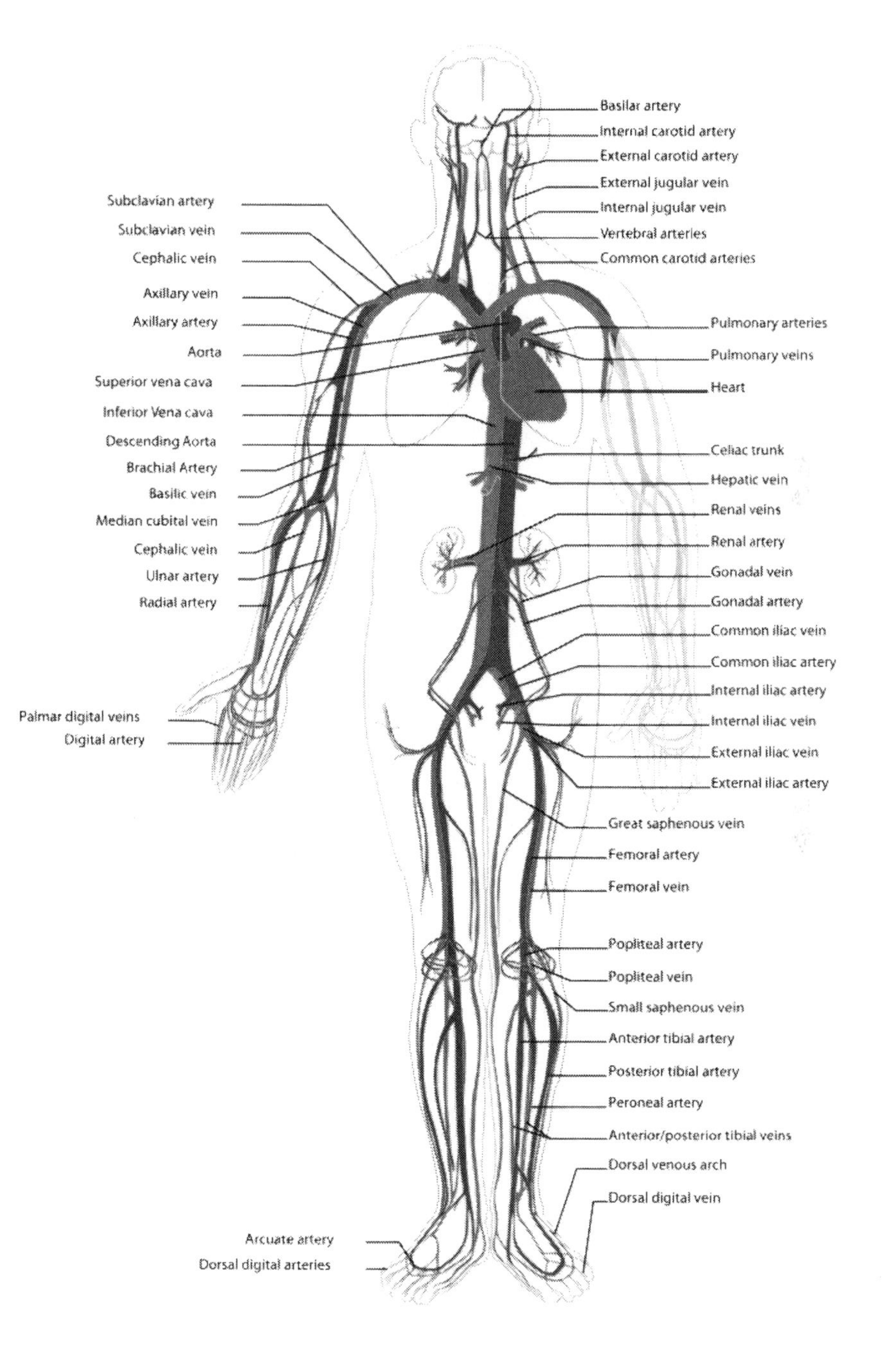
Basilar artery
Internal carotid artery
External carotid artery
External jugular vein
Internal jugular vein
Vertebral arteries
Common carotid arteries
Subclavian artery
Subclavian vein
Cephalic vein
Axillary vein
Axillary artery
Aorta
Superior vena cava
Inferior Vena cava
Descending Aorta
Brachial Artery
Basilic vein
Median cubital vein
Cephalic vein
Ulnar artery
Radial artery
Pulmonary arteries
Pulmonary veins
Heart
Celiac trunk
Hepatic vein
Renal veins
Renal artery
Gonadal vein
Gonadal artery
Common iliac vein
Common iliac artery
Internal iliac artery
Internal iliac vein
External iliac vein
External iliac artery
Palmar digital veins
Digital artery
Great saphenous vein
Femoral artery
Femoral vein
Popliteal artery
Popliteal vein
Small saphenous vein
Anterior tibial artery
Posterior tibial artery
Peroneal artery
Anterior/posterior tibial veins
Dorsal venous arch
Dorsal digital vein
Arcuate artery
Dorsal digital arteries

Circulation

There are three means of circulation, which transport blood around the body.

Pulmonary Circulation:

Blood picks up oxygen from the lungs and releases carbon dioxide.

1. Blood that is not oxygenated moves from the right ventricle into the pulmonary arteries.
2. The blood travels through arteries and arterioles that are progressively smaller into the capillaries of the lungs.
3. Once the blood reaches the alveoli, it exchanges carbon dioxide with oxygen.
4. Blood that is oxygenated returns through venules and veins into the pulmonary veins. The blood is then carried back into the left atrium.

Systemic Circulation: The left ventricle pumps blood that carries oxygen and nutrients to cells throughout the body. It also transports waste products that need to be excreted.

The aorta branches out into vessels that provide for certain body organs and areas. The left *common carotid artery* provides blood to the brain. The *left subclavian artery* provides blood to the arms. The *innominate artery* provides blood to the upper chest. As it moves through the abdomen and thorax, the aorta provides blood to the organs of the genitourinary and GI systems, lower chest, abdominals, and spinal column. The aorta splits into *iliac arteries* and then *femoral arteries.*

Perfusion

The number of vessels increase significantly as the arteries divide into smaller units. This increases the tissue area that blood flows to, which is called the *area of perfusion*.

Dialation

Sphincters control the blood that flows into the tissues. The sphincters dilate, which allows more blood flow when it is required - or they close to prevent blood flow to other areas - or they constrict to cause an increase in the blood pressure.

Low Pressure Area

The capillary bed provides blood to the largest number of cells, and the capillary pressure is very low which permits nutrient, oxygen, and carbon dioxide exchange with body cells. The blood flows from the capillaries into the venules, and then into the veins.

Backflow Prevention

The backflow of blood is prevented by valves in the veins. Any pooled blood in valved segments is transported towards the heart as a result of pressure from the moving blood volume below. The veins form two branches: the *superior vena cava* and *inferior vena cava* – these transport blood back to the right atrium.

Coronary Circulation:

The heart depends upon the coronary arteries for its fresh supply of oxygenated blood. It also relies of the cardiac veins to remove the blood that has been depleted of oxygen.

Blood is ejected from the left ventricle into the aorta during systole. Blood moves out of the heart and within the coronary arteries during diastole. This nourishes the heart.

The right coronary artery provides blood to the right atrium, some of the left atrium, the majority of the right ventricle, and the lower part of the left ventricle.

The left coronary artery provides blood to the left atrium, the majority of the left ventricle, and the majority of the interventricular septum.

The arterial system of blood vessels provides oxygenated blood to the heart. They also provide blood to the venous system that removes blood that has been depleted of oxygen. Below is an illustration of the coronary circulation system.

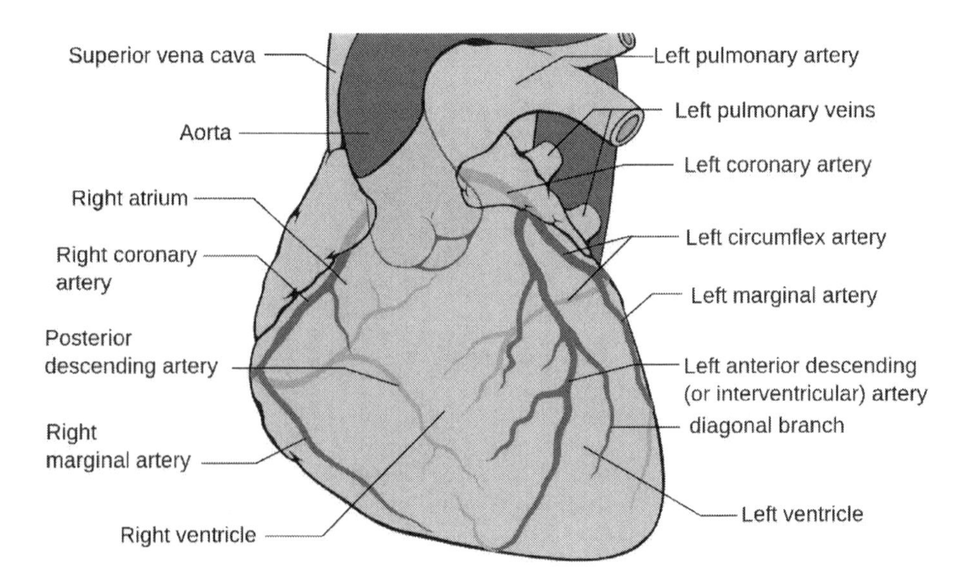

Section 10: The Hematologic System

The hematologic system is made up of both blood and bone marrow. Blood has the responsibility of delivering oxygen and nutrients to tissues, removing wastes, and the transportation of gasses, immune cells, blood cells, and hormones in the body.

Hematopoiesis: this is a process through which the hematologic system manufactures new blood cells.

There are *multipotential stem cells* within bone marrow, which include five different cell types that are named *unipotential stem cells.* The four types of unipotential stem cells are as follows:

- Erythrocyte
- Granulocyte
- Agranulocyte
- Platelet

Below is an illustration of the process – from the birth of unipotential stem cells, to when they reach fully formed cells:

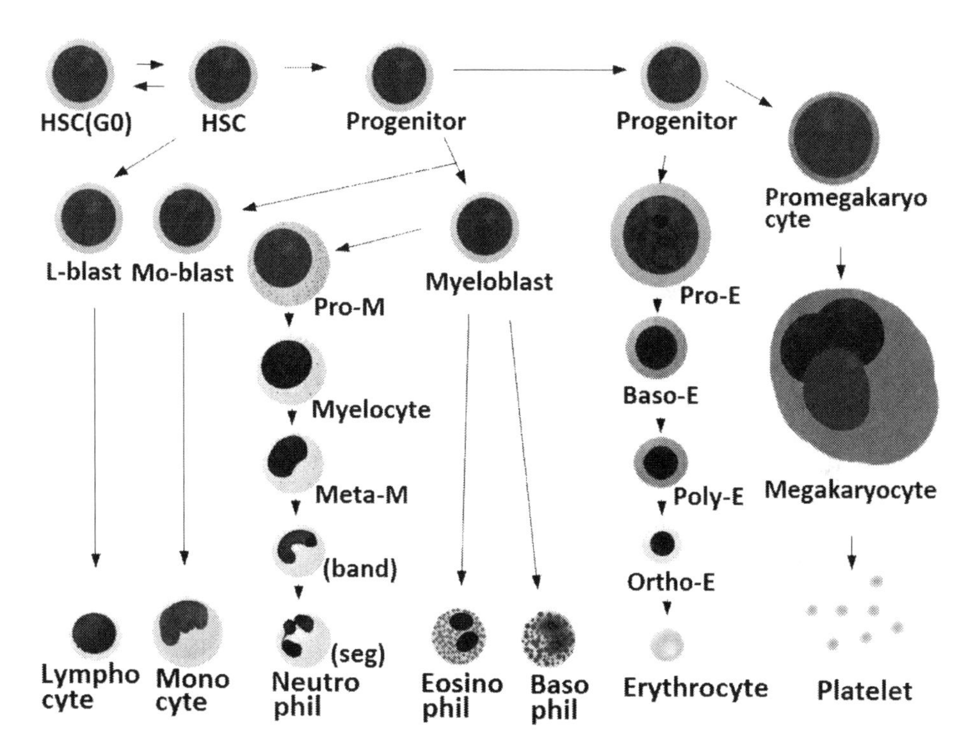

The Components of Blood

Blood is made up of a number of formed elements/blood cells; these are held in a fluid that is called *plasma.*

1. Red Blood Cells (Erythrocytes)

Red blood cells are responsible for the transportation of both oxygen and carbon dioxide in and out of the tissues of the body. Within the red blood cells is *hemoglobin,* which is a substance that carries oxygen and gives blood its red color.

The average life span of red blood cells is 120 days. When red blood cells are released into circulation by bone marrow, they are in immature form called *reticulocytes.* This immature form generally takes one day to mature into red blood cells. The spleen functions by removing old red blood cells from circulation.

The volume of reticulocytes that are releases is generally equal to that of the removal of the old red blood cells. When there is a depletion of red blood cells, there is an increase in the production of reticulocyte production by the bone marrow. This maintains the normal red blood cell count.

2. White Blood Cells

White blood cells function in the defense and immune systems in the body. There are five types of white blood cells:

Granulocytes

Granulocytes consist of *neutrophils, eosinophils,* and *basophils* - these are *polymorphonuclear leukocytes.* Each type of cell has different characteristics and is activated by different stimuli.

Neutrophils make up 50 to 70 percent of white blood cells that are in circulation. These are *phagocytic* cells that ingest and digest materials at infections sites. They pass through capillary walls to arrive at injury sites (*diapedesis*).

Eosinophils make up 0.3 to 7 percent of white blood cells in circulation. Like neutrophils, they also leave the bloodstream via diapedesis. They are released as a response to allergic reactions, and function by ingesting antigen-antibody complexes in loose connective tissue.

Basophils generally make up less than two percent of white blood cells in circulation. They have very little to no ability to ingest foreign or harmful materials (phagocytic ability). They secrete histamine that allows fluids to easily pass from the capillaries into the tissues of the body.

Agranulocytes

Agranulocytes, such as *monocytes* and *lymphocytes,* have nuclei without lobes and do not have specific cytoplasmic granules.

Monocytes are the largest white blood cells that make up 1 to 9 percent of white blood cells in circulation. They enter the tissues via diapedesis, and they mature into tissue *macrophages* outside of the bloodstream. Macrophages are able to travel through the body in response to inflammation, and they are part of the *reticuloendothelial system* that provides the body with defense against infection and the disposal of cell breakdown material. Macrophages are contained in structures such as the liver, lymph nodes, and spleen to provide defense against the invasion of organisms - these are structures that filter significant amounts of body fluid. They are phagocytes that ingest depleted neutrophils (cellular debris), microorganisms, and necrotic tissue. A key function of macrophages is to stimulate the healing of wounds.

Lymphocytes are the smallest white blood cells that make up 20 to 43 percent of white blood cells in circulation. They arise from stem cells within bone marrow. The three types of lymphocytes are:

- **T lymphocytes:** these attack an infected cell directly.
- **B lymphocytes:** these function in the production of antibodies against antigens.
- **Natural killer cells:** these provide surveillance for the

immune system and resistance to infection.

Comparing Granulocytes and Agranulocytes

Granulocytes, including neutrophils, basophils, and eosinophils are the first to step up to defend against invading organisms. Following this, agranulocytes, including monocytes and lymphocytes, generally move around freely on 'surveillance' when there is inflammation. They will infiltrate further into the structures with large amounts of fluid to provide defense against invading organisms.

3. Platelets

Platelets are colorless, small, disks that are cytoplasmic fragments, which arise from megakaryocytes. These are cells in the bone marrow. They have a life span of about ten days and have three main functions:

- Stimulating the contraction of blood vessels that have been damaged. This minimizes the loss of blood.
- The formation of hemostatic plugs within blood cells that have been injured.
- Providing the materials that speed up the coagulation of blood.

Blood Clotting (Hemostasis)

Hemostasis or blood clotting is a process in which platelets, plasmas, and coagulation factors cooperate to control bleeding. When there is a rupture of a blood vessel, at the site of the injury there is a reduction in the caliber of blood vessels (local vasoconstriction), along with the clumping of platelets. This provides the initial prevention of hemorrhage. Thromboplastin is then released by the damaged cells and this activates the coagulation system's *extrinsic pathway*.

For a stable clot to form, a complex clotting mechanism is required, this is called the *intrinsic pathway*. A protein (factor XII) activates the clotting mechanism. It arises from plasma and tissue and it is one of the 12 substances needed for coagulation. The final stage of the clotting mechanism is a *fibrin clot* - this is a insoluble protein that is found at the location of the injury.

Both platelets and plasma provide materials that work alongside coagulation factors. These serve as precursor compounds in the coagulation (clotting) of the blood.

12 Coagulation Factors

The coagulation factors work in a 'chain-reaction' process whereby one factor activates the next factor. They are ordered in roman numerals as follows:

- **Factor I - Fibrinogen:** this is a glycoprotein that is synthesized in the liver and during the coagulation cascade is converted into *fibrin*.
- **Factor II - Prothrombin:** this is a protein that is produced in the liver and during coagulation it is converted to thrombin.
- **Factor III - Thromboplastin (Tissue Factor):** this is needed to activate the extrinsic pathway and it is released from tissue that is has been damaged.
- **Factor IV - Calcium Ions:** these are necessary throughout the whole clotting process.
- **Factor V - Proaccerlerin (Labile factor):** this is a protein that is produced in the liver. It functions throughout the common pathway phase of the clotting system.
- **Factor VII - Proconvertin (Stable Factor):** this is a protein that is produced in the liver. It is activated by factor III

within the extrinsic system.

- **Factor VIII - Antihemophilic Factor:** this is a protein that is produced in the liver and is needed during the intrinsic stage of the clotting system.
- **Factor IX – Plasma Thromboplastin:** this is a protein produced in the liver. It is needed during the intrinsic stage of the coagulation system.
- **Factor X – Stuart Factor:** this is a protein produced in the liver and it is needed in the coagulation system's common pathway.
- **Factor XI – Plasma Thromboplastin Antecedent:** this is a protein that is produced in the liver and is needed during the intrinsic stage of the coagulation system.
- **Factor XII – Hageman Factor:** this is needed in the intrinsic pathway.
- **Factor XIII – Fibrin Stabilizing:** this is a protein that is needed to provide stabilization for the fibrin strands within the common pathway phase of the clotting system.

There are three interconnected process that happen when a blood vessel is injured.

Constriction and Aggregation

Following an injury, the vessels that are damaged will immediately contract. This is called *constriction,* which reduces blood flow. Along with this, collagen of the impacted cells stimulates the platelets and begins to cluster together. This is called *aggregation,* which seals the injury site temporarily.

Clotting Pathways

The change of blood from a liquid to a solid during the coagulation process is instigated via two separate pathways:

- **Intrinsic Pathway:** this is stimulated when plasma makes contact with the surfaces of the affected vessels.
- **Extrinsic Pathway:** this is stimulated when tissue factor makes contact with one of the coagulation factors.

Blood Groups

Antigens or glycoproteins on the surface of the red blood cells determine blood groups. The most significant blood antigens are A, B, and Rh.

ABO Groups:

The most important system for the classification of blood is the testing for the presence of A and B antigens on red blood cells.

- **Type A Blood:** this has A antigen on its surface.
- **Type B Blood:** this has B antigen.
- **Type AB Blood:** this has A and B antigens.
- **Type O Blood:** this does not have A or B antigens.

Antibodies in plasma interact with A and B antigens. This results in the cells combining into a mass - the cells *agglutinate*. Plasma is unable to have antibodies to its own cell antigen, therefore blood that is type A has A antigen and does not have A antibodies, however it does have B antibodies.

Crossmatching

Accurate blood *crossmatching* is essential, particularly for blood transfusions. When blood is being transferred from a donor to a recipient, it must be compatible. Here are the blood groups that are compatible:

- **Type A Blood:** is compatible with type A or O blood.
- **Type B Blood:** compatible with type B or O blood.
- **Type AB Blood:** compatible with type A, B, AB, and O blood.
- **Type O Blood:** only compatible with type O blood.

Blood Group	Antigens	Antibodies	Can give blood (RBC) to	Can receive blood (RBC) from
AB	A and B	None	AB	AB, A, B, O
A	A	B	A and AB	A and O
B	B	A	B and AB	B and O
O	None	A and B	AB, A, B, O	O

Section 11: The Lymphatic System and Immunity

The immune system of the body serves the function of providing defense against the attack of harmful organisms and chemical toxins. Both the organs and tissues of the immune system are named *lymphoid*. This is because they are all involved with the development, and distribution of *lymphocytes,* which are a type of white blood cell. There are three main components of the immune system:

- The ***central lymphoid*** organs and tissue.
- The ***peripheral lymphoid*** organs and tissue.
- The ***accessory lymphoid*** organs and tissue.

The immune system and the blood are closely related. Their cells both originate in the bone marrow, and the immune system utilizes the bloodstream to send cells to an invasion site.

The Organs and Tissues of the Immune System

The immune system is made up of organs and tissues where the lymphocytes are found, along with cells that circulate in the bloodstream. The organs and tissues are central, peripheral, or accessory.

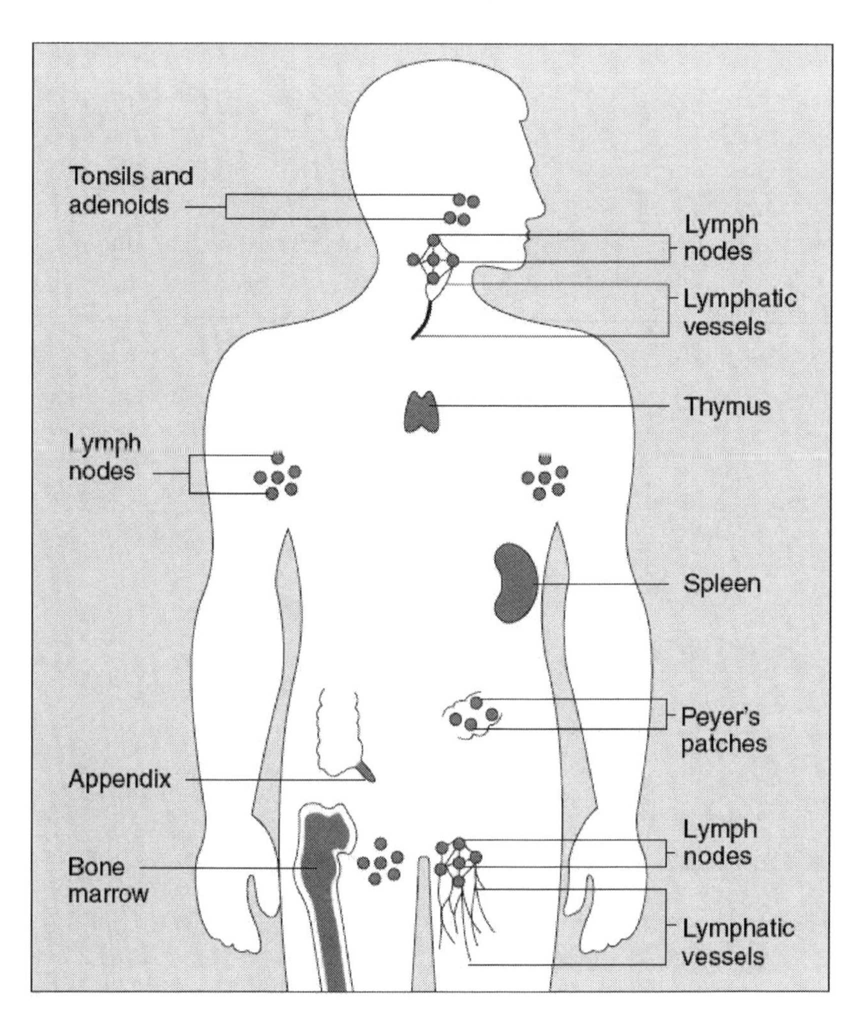

The Central Lymphoid Organs and Tissues

B cells and T cells are the two main types of lymphocytes - both the bone marrow and thymus play a role in the development of these cells.

Bone Marrow

Within the bone marrow are stem cells, which can mature in a number of different types of cells. These cells are referred to as *multipotential,* which means they are able to take a variety of forms. *Hematopoiesis* is the process in which the immune system cells and blood cells develop from stem cells.

Following their differentiation from other stem cells, a number of the cells that will become immune system cells act as sources for lymphocytes. Other cells of this group will progress into cells that ingest microorganisms called phagocytes. The cells that become lymphocytes becoming more differentiated by developing into ***B cells*** that develop in the bone marrow, or ***T Cells*** that develop in the thymus.

B cells and T cells are spread throughout the lymphoid organs, particularly in the lymph nodes and spleen. The B and T lymphocytes have specialized receptors, which respond to certain shapes of antigen molecule. The receptor in B cells is *immunoglobulin,* also named an *antibody*. These attack invading pathogens and instruct other cells to attack on their behalf.

The Thymus

In babies and infants, the *thymus* is a mass of lymphoid tissue that is positioned over the heart's base in the mediastinum. Following birth, the thymus assists in the formation of T cells. However, after this the thymus does not play any role in the immunity for the body. It develops up until the age of puberty and then starts to degenerate until there is just a small residual left.

A process called 'T-cell education' occurs in the thymus. The cells are 'taught' to identify *self cells,* which are cells from the same body, and to differentiate these from *nonself cells*. There are a number of T cells and each serve a specific function:

- Memory T cells
- T4 cells, or helper cells.
- T8 cells, or regulatory cells.
- Cytotoxic T cells, or natural killer cells.

The Peripheral Lymphoid Organs and Tissues

The lymph nodes, lymphatic vessels, and spleen make up the peripheral structures of the immune system.

Lymph Nodes

The lymph nodes are oval-shaped formations that are found in a network of lymph channels. The lymph nodes are most prevalent in the neck, head, axillae, pelvis, abdomen, and groin. They assist in the removal and destroying of antigens, which move around in the blood and lymph. A fibrous capsule encloses each lymph node, and connective tissue extends from this into the node, dividing it into the following three sections:

- **The Superficial Cortex**: this contains follicles that are made up of mostly B cells.
- **The Deep Cortex and Interfollicular:** this consists of predominantly T cells.
- **The Medulla:** this consists of a number of plasma cells, which secrete immunoglobulins.

Lymphatic Vessels

The lymphatic vessels are a system of thin-walled drainage channels called *lymphatic vessels,* which connect the lymphatic tissues. The *afferent lymphatic vessels* serve the function of carrying lymph fluid into lymph nodes. The lymph gradually travels though the node and it is collected by the *efferent lymphatic vessels.*

Spleen

The spleen is an oval, dark red structure, which is the largest lymphatic organ. It is positioned below the diaphragm in the upper left portion of abdomen. Connective tissue, which comes from the fibrous capsule around the spleen, spans into the interior of the spleen. The interior of the spleen is called the *splenic pulp,* which contains white and red pulp. White pulp is made up of lymphocytes that surround branches of the splenic artery. Red pulp is made of a system of blood-filled sinusoids, which are reinforced by mononuclear phagocytes and reticular fibers, as well as lymphocytes, monocytes, and plasma cells.

Functions of the Spleen:

- Breaking down and engulfing depleted red blood cells with phagocytes, which causes the release of hemoglobin.
- Filtering and removing bacteria, along with other foreign substances that enter the bloodstream.
- Interacting with lymphocytes to produce an immune system response.
- Storing blood and 20-30% platelets.

The Accessory Lymphoid Organs and Tissues

The accessory lymphoid organs and tissues are made up of the tonsils, adenoids, appendix, and Peyer's patches. They remove foreign substances in a similar way to that of the lymph nodes. These lymphoid organs and tissues are found in areas where microbial prevalence is more likely. Such areas are the tonsils and adenoids (*nasopharynx*), and the appendix and Peyer's patches in the abdomen.

The Immune System Function

The body's ability to defend against invading organisms and toxins is referred to as *immunity*. The primary purpose of this defense is to prevent damage to the tissues and organs, by identifying and eliminating antigens (bacteria, viruses, fungi, parasites).

There are three main approaches the immune system takes to perform its functions effectively: ***protective surface phenomena***, ***general host defenses***, and ***specific immune responses***.

Protective Surface Phenomena

The entry of harmful organisms is prevented by physical, mechanical, and chemical barriers. The first line of defense is provided by mucous membranes and intact and healing skin. Low pH and skin *desquamation* further prevent bacterial build up. Seromucous surfaces are defended with antibacterial substances such as the enzyme ***lysozyme***. This is found in saliva, tears, and nasal secretions.

Within the respiratory system, nasal hairs and turbulent airflow filter out foreign substances. Immunoglobulin within nasal secretion deters the adherence of microbes, and the mucous layer provides further protection.

Within the gastrointestinal (GI) tract, bacteria are removed by saliva, peristalsis, swallowing, and defecation. On top of this, the gastric secretions that are low pH are bactericidal, meaning they kill bacteria. This makes the stomach essentially free from all live bacteria. The rest of the GI system is protected by colonization resistance, whereby bacteria that is already present stops other microorganisms from settling.

Within the urinary system, low urine pH, urine flow, and immunoglobulin work in unison to prevent the colonization of bacteria. In men, prostatic fluid also serves the function of killing bacteria.

General Host Defenses

In order to recognize and remove invading antigens, the immune system produces nonspecific cellular responses when an antigen enters the skin or mucous membrane. The *inflammatory response* is the first nonspecific response for an antigen. This process includes vascular and cellular changes, which involve the release of chemicals such as heparin, kinin, and histamine. These changes get rid of microorganisms, dead tissue, and toxins.

An example of the inflammatory response is illustrated below:

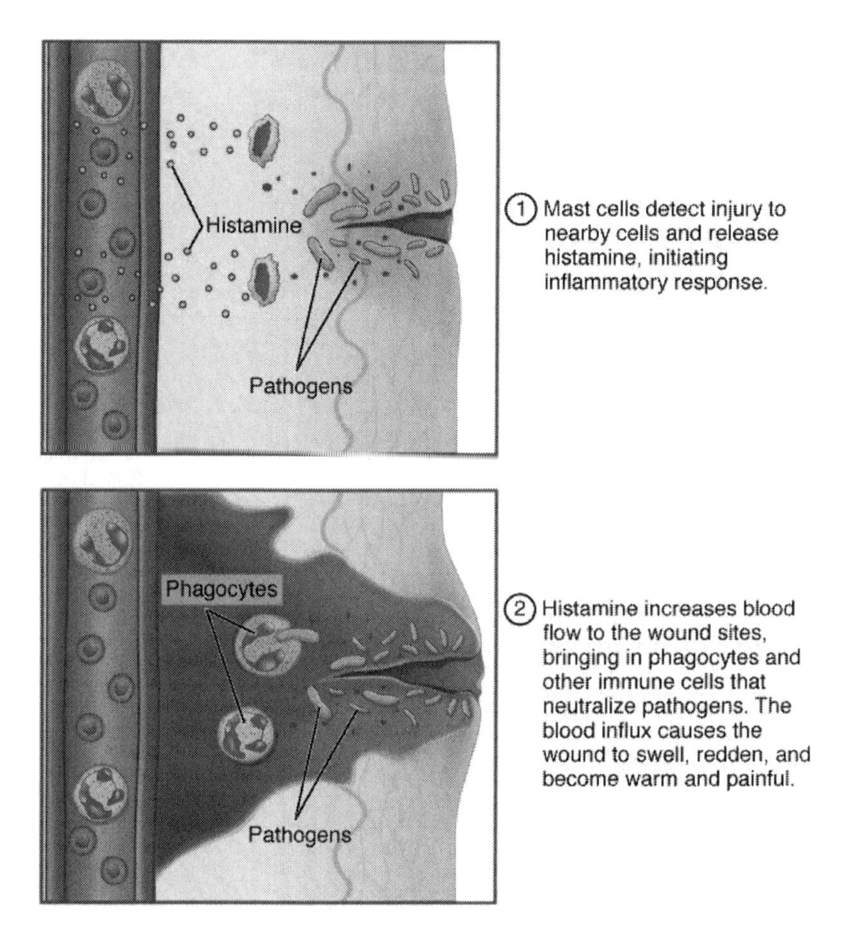

Specific Immune Responses

All foreign substances that enter the body give rise to the same general host defenses. However, some microorganisms or molecules in particular can stimulate specific immune responses which can include a specialized immune cells. The specific responses are classed as either humoral immunity or cell-mediated immunity, and they are produced by B cells and T cells.

Humoral Immunity

In the humoral immunity response, an antigen causes B cells to split up and change into plasma cells. The plasma cells secrete antigen-specific immunoglobulins into the bloodstream. There are five types of immunoglobulins:

- **IgA, gM, and IgG:** these provide protection against viral and bacteria attacks.
- **IgD:** this serves as a B cell antigen receptor.
- **IgE:** this produces an allergic response.

Immunoglobulins have a Y shaped molecular structure. The top of the fork of the Y attaches to a particular antigen, and the lower stem of the Y allows is immunoglobulin to combine with other structures within the immune system. Immunoglobulins work in a number of ways, depending upon the antigen:

- Immunoglobulins called antitoxins link with toxins that bacteria produce, thereby disabling certain bacteria.
- They are able to coat or opsonize bacteria, thereby making them available for removal by phagocytosis.
- They can link to antigens, which causes *complement* enzymes to be produced and circulated by the immune system.

The Sequence of Response

Following an exposure to an antigen, there is a time lag that takes place in which there is little to no antibody detected. Throughout this period, the B cell identifies the antigen, along with the sequence of division and differentiation. Following this, the formation of antibodies begins.

The First Response: the *primary antibody response* takes place 4-10 days after the exposure of the antigen. The levels of immunoglobulin rise and then rapidly dissipate. There is the formation of IgM antibodies.

The Second Response: a repeated exposure to the antigen produces a *secondary antibody response*. Memory B cells create antibodies, which are mostly IgG, and achieve their peak levels in 1-2 days. These levels remain at this level for a number of months and then gradually fall.

Antigen-Antibody Complexes: an antigen-antibody complex forms after the antibody reacts to the antigen, and this has a number of functions. Firstly, the antigen is processed by a macrophage and is presented to antigen-specific B cells. The antibody then stimulates the complement system, which destroys the antigen via *enzymatic cascade*.

Complement System: this is activated by antigen-antibody reactions or a tissue injury. It joins humoral and cell-mediated immunity, along with attracting phagocytic neutrophils and macrophages to the site of the antigen. The complement system is made up of around 25 enzymes that supplement the work of antibodies by helping with destroying bacteria cells or aiding *phagocytosis*.

Phagocytosis

Phagocytosis is a defense mechanism conducted by macrophages and neutrophils, which removes antigens and microorganisms that invade the skin. Here is the step-by-step process of how macrophages achieve phagocytosis:

1. **Chemotaxis:** The chemotactic factors attract macrophages to the site of the antigen.
2. **Opsonization:** Immunoglobin G (the antibody) or complement fragment coats the microorganism, which enhances macrophage attaching to the antigen. This is now

called an opsinogen.

3. **Ingestion:** the membrane of the macrophage is extended around the opsonized microorganism, which engulfs it within a phagosom (vacuole).
4. **Digestion:** the phagosome joins with the lysosomes as it moves away from the cell edge. This forms a phagolysosome, in which the destruction of antigen occurs.
5. **Release:** Following digestion, the digestive debris is expelled by the digestive debris, which includes lysosomes, complement components, prostaglandins, and interferon.

The Complement Cascade

A ripple motion is set into effect after a complement substance being triggered as a result of an antibody being joined with an antigen. The activation of each component results in the acting on the next component in the sequence. This series of controlled steps is called the *complement cascade*. The cascade creates the *membrane attack complex,* in which fluid and molecules flow in and out. Following this, the target cell expands and bursts. The complement cascade also creates the inflammatory response, the attraction of neutrophils, and the coating of target cells.

Cell-Mediated Immunity

Cell-mediated immunity provides protection for the body against, bacterial, viral, and fungal infections. It does this by disabling the antigen and gives resistance against both transplanted cells and tumor cells. During this response, the antigen is processed by a macrophage and then given to T cells. Some of the T cells kill the antigen, and others produce *lymphokines* that stimulate macrophages that kill the antigen.

Section 12: The Respiratory System

The respiratory system is responsible for exchanging oxygen and carbon dioxide in the lungs and body tissues. Along with this, it assists in the regulation of the acid-base balance in the body. The respiratory system is made up of a respiratory zone and a conducting zone.

Respiratory Zone: Made up of the bronchioles, alveoli, alveolar ducts - it performs the exchange of gas.

Conducting Zone: Made up of the continuous passageway (nose, pharynx, trachea, larynx, bronchi, bronchioles, and terminal bronchioles). This transports air into and out of the lungs.

The respiratory system is made up of the upper respiratory tract, the lower respiratory tract, and the thoracic cavity.

The Upper Respiratory Tract

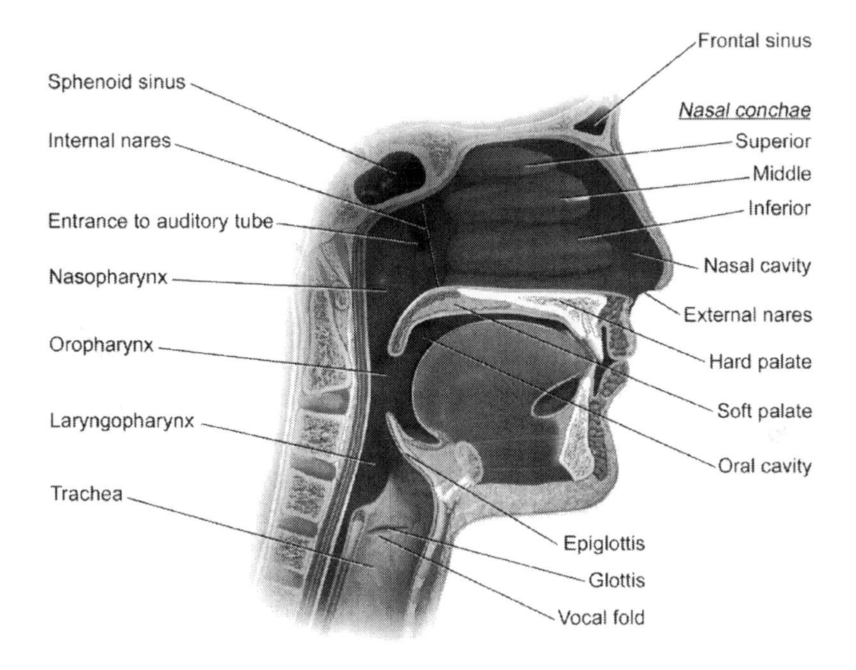

The upper respiratory tract is made up of the nostrils and nasal passages, mouth, oropharynx, nasopharynx, laryngopharynx, and larynx. They function by filtering, warming, and humidifying air that is inspired. Along with this, they detect both taste and smell.

Nasal Passages and Nostrils

Air comes into the body through the notrils, and during this the vibrissae (small hairs) function by filtering out dust and foreign material. Following this, the air then enters into the two nasal passages that are separated by the septum. The anterior walls of the nasal passages are formed by cartilage, and the posterior walls are formed by conchae/turbinates.

Air is warmed and humidified by the conchae prior to entering the nasopharynx. A layer of mucus catches foreign particles, which are carried to the pharynx by the cilia. These foreign particles will then be swallowed.

The Sinuses and Nasopharynx

There are four paranasal sinuses that are positioned in the frontal, maxillary, and sphenoid bones. Air travels from the nasal cavity and enters into the nasopharynx via the choanae. The choanae are constantly open. The nasopharynx is positioned above the throat and behind the nose.

Oropharynx, Laryngopharynx, Larynx

The posterior wall of the mouth is called the oropharynx, which links the nasopharynx and the laryngopharynx. The laryngopharynx reaches to the larynx and the esophagus.

The vocal cords are located in the larynx, and it links the pharynx with the trachea. The walls of the larynx are formed by muscles and cartilage.

The Lower Respiratory Tract

The lower respiratory tract is made up of the trachea, bronchi, and lungs. Lining the lower tract is a mucous membrane, which has hairlike cilia that continuously clean the tract and transport foreign matter to be swallowed.

Trachea and Bronchi

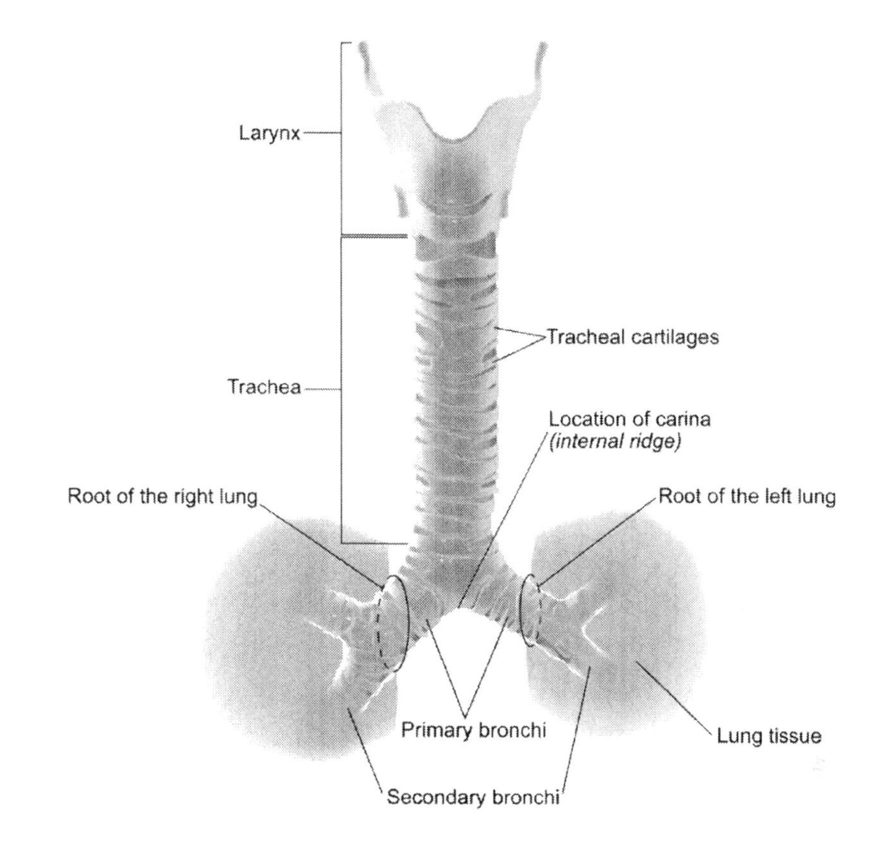

The trachea spans from the cricoid cartilage to the carina and it is protected by cartilage rings.

The bronchi start at the carina. The right bronchus provides air to the right lung, and the left bronchus provides air to the left lung. The secondary bronchi enter the lungs and the pleural cavities and the hilum, which is a slit of the lung's surface.

Each lobar bronchus moves into a lobe in each of the lungs. Each of the lobar bronchi divides into tertiary bronchi. These are segmental bronchi that carry on to branch out into smaller bronchi, and eventually branch into bronchioles. When the bronchi are large, they are made up of cartilage, epithelium, and smooth muscle. As they get smaller they lose the cartilage/smooth muscle and are solely made up of epithelial cells.

Respiratory Bronchioles

The acinus and terminal bronchioles are in each bronchiole. The acinus is the main gas exchange respiratory unit, and the terminal bronchioles branch into smaller respiratory bronchioles within the acinus. The respiratory bronchioles connect directly in the alveoli.

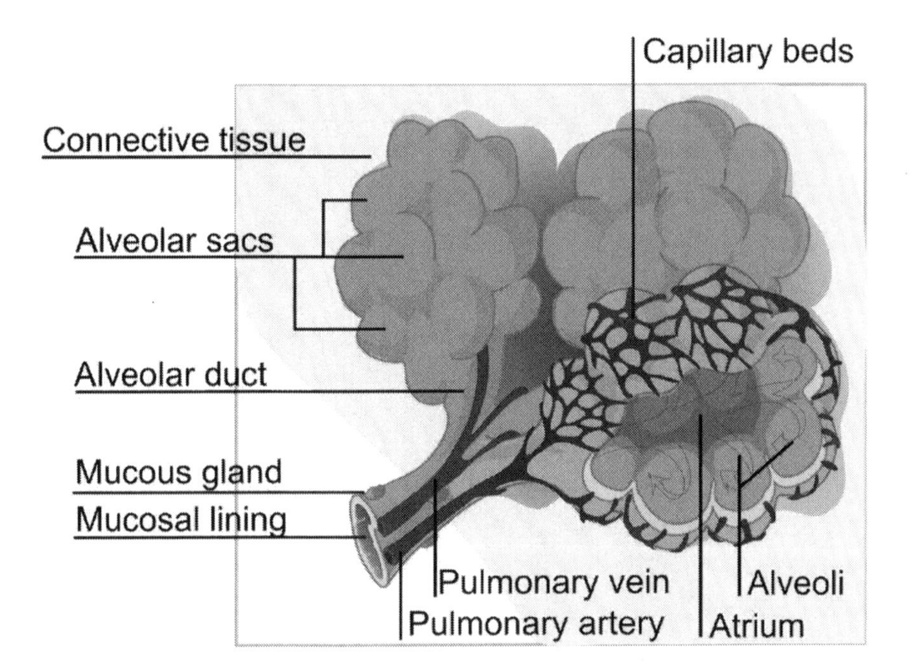

The respiratory unit is made up of the respiratory bronchiole, alveolar sac, alveolar ducts, and alveoli. Within the alveoli, oxygen that is inhaled is diffused into the blood, and carbon dioxide is infused into the air from the blood. This is called *gas exchange.*

The bronchioles turn into alveolar ducts, which then become alveolar sacs. The walls of the alveolar are comprised of two types of cells:

- **Type I Cells:** this is where the gas exchange takes place. They are thin and flat cells.
- **Type II Cells:** these are cells that secrete a substance called *surfactant*. This aids the process of gas exchange by lowering the tension on the surface.

The Lungs

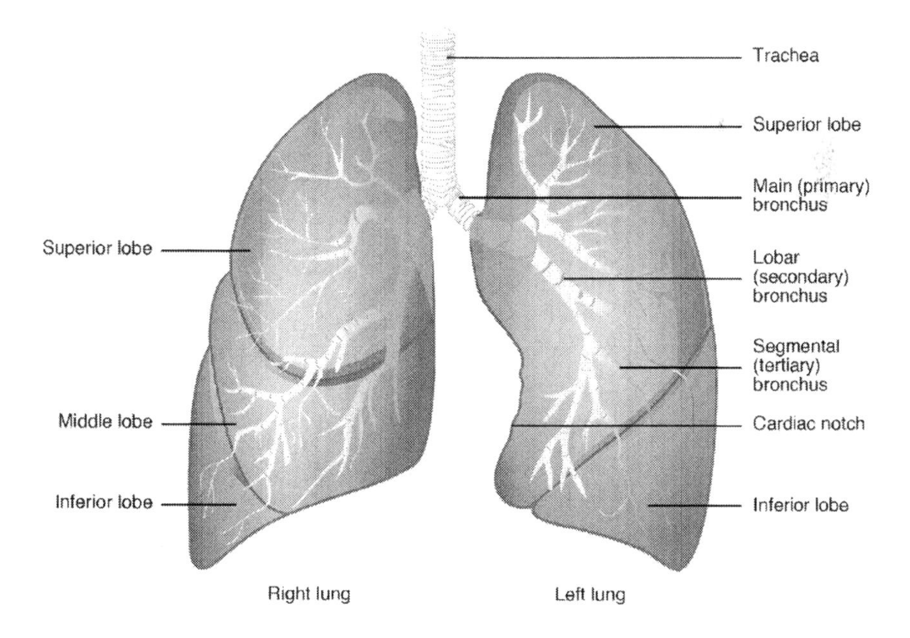

The lungs are cone-shaped and are positioned in the right and left pleural cavities. They wrap around the heart and are secured by root and pulmonary ligaments. The right lung is larger than the left - it contains three lobes and is responsible for 55 percent of the gas exchange. The left lung is made up of two lobes. The concave bases of both lungs sit on top of the diaphragm.

The Pleura and the Pleural Cavities

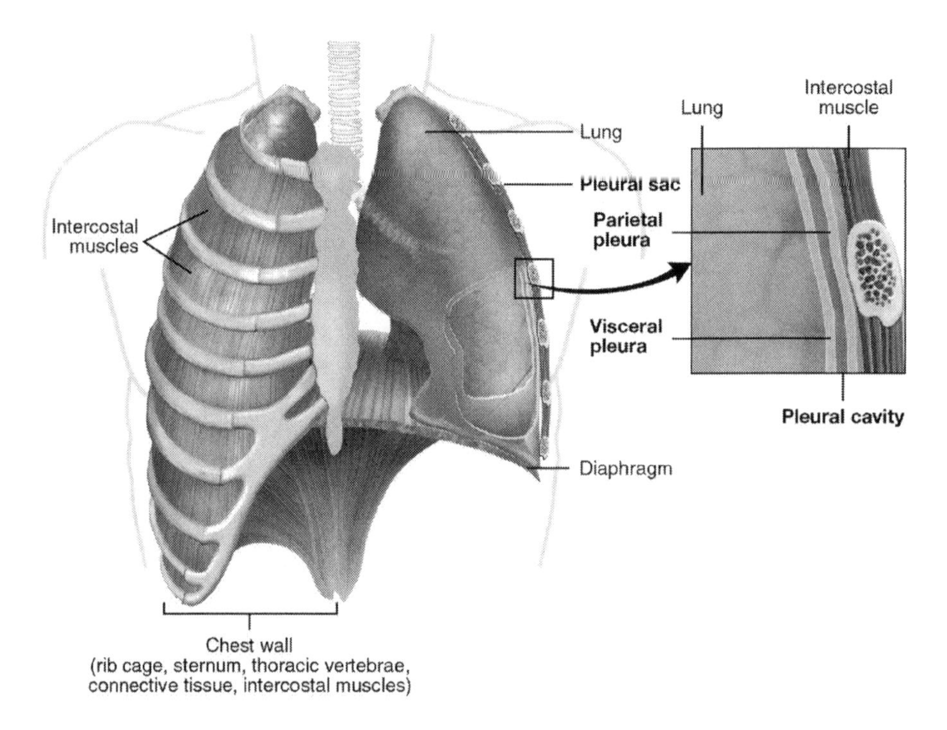

The *pleura* is the membrane that entirely surrounds the lung, and it is made up of a parietal layer and a visceral layer.

- **Visceral Pleura:** this wraps around the whole surface of the lung and the areas in between the lobes.
- **Parietal Pleura:** this provides lining for the chest wall's inner surface and the diaphragm's upper surface.

The pleural cavity is the small area in between the visceral and parietal layers, and it contains a thin layer of *serous fluid*. The serous fluid serves two main functions:

- Lubrication of the pleural surfaces. As the lungs expand and contract, this allows for them to slide smoothly against each other.
- It produces a connection between the layers, causing the lungs to move along with the chest wall whilst breathing.

The Thoracic Cavity

The thoracic cavity is the area that is surrounded by the following:

- Diaphragm
- Scalene muscles
- Fasciae of the neck
- Ribs
- Intercostal muscles
- Vertebrae
- Ligaments
- Sternum

The Mediastinum

The mediastinum is the space between the lungs, which contains the:

- Heart and pericardium
- Thoracic aorta
- Pulmonary artery and veins
- Azygos veins and venae cavae
- Lymph nodes, thymus, and vessels
- Vagus, cardiac, and phrenic nerves.
- Trachea, esophagus, and thoracic duct

The Thoracic Cage

The thoracic cage is made up of bone and cartilage, and it provides support and protection for the lungs. This allows the lungs to expand and contract.

- **The Posterior Thoracic Cage:** The posterior thoracic cage is made up by the vertebral column and the 12 pairs of ribs.
- **The Anterior Thoracic Cage:** The anterior thoracic cage is made up of the sternum, manubrium, ribs, and xipoid process. It provides protection for the mediastinal organs that are located in between the pleural cavities.

The Ribs

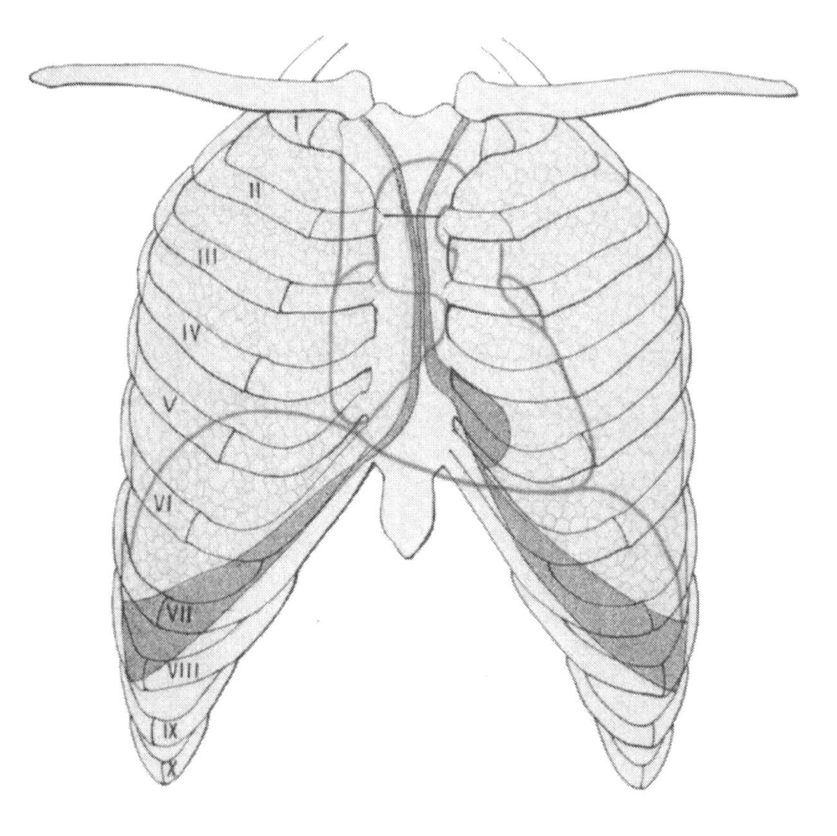

Ribs 1 to 7: these ribs are attached directly to the sternum.

Ribs 8 to 10: these ribs are attached to the cartilage of the previous ribs.

The remaining pairs of ribs: these ribs float freely and are not attached to the anterior thoracic cage.

Inspiration and Expiration

The breathing process includes two actions: an active process called *inspiration* and a passive process called *expiration*. The actions are dependent upon the respiratory muscle function, along with the differences in pressure in the lungs.

The external intercostal muscles assist the diaphragm during respiration. The diaphragm is the main muscle of respiration, and it lengthens the chest cavity. The external intercostal muscles contract and this expands the anteroposterior diameter. These actions produce a decrease in intrapleural pressure, which causes inspiration to occur.

The expiration happens as a result of an intrapleural pressure increase, and this occurs due to the diaphragm rising and the intercostal muscles relaxing.

The respiration muscles assist the chest cavity to expand and then contract. Air movement is produced by the differences in pressure between the lungs and atmospheric air. The diagram below illustrates the different muscles that work to produce inspiration and expiration.

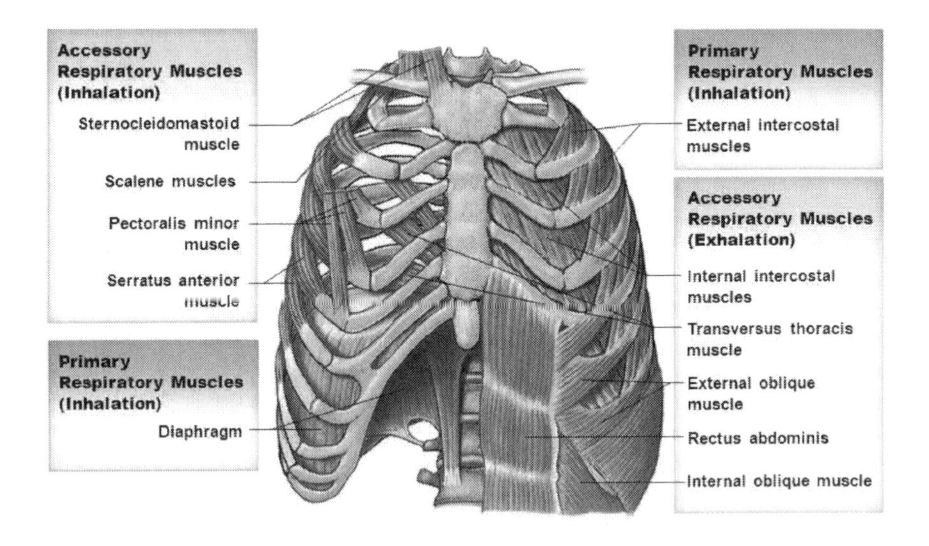

Forced Inspiration

The accessory muscles of respiration contribute when the body requires an increased amount of oxygenation. An example of this is during exercise or a disease state that requires forced inspiration and active expiration.

The following occurs during forced inspiration:

- The upper chest's pectoral muscles raise the chest, which increases the anteroposterior diameter.
- The sternum is raised by the sternocleidomastoid muscles in the neck.
- The upper chest is elevated and expanded by the scalene muscles.
- The thoracic cage is raised by the posterior trapezius muscles in the upper back.

Active Expiration

Throughout active expiration, the chest's transverse diameter is shortened by the intercostal muscles. Along with this, the lower chest is pulled down by the abdominal rectus muscles, which depresses the lower ribs.

Internal Respiration and External Respiration

Effective external respiration happens as a result of gas exchange in the lungs. Internal respiration is gas exchange that takes place in the tissues by diffusion. External respiration takes place through the following processes:

- **Ventilation**: gas delivery in and out of the pulmonary airways.
- **Pulmonary perfusion**: blood flowing through the right side of the heart, then through the pulmonary circulation, and into the heart's left side.
- **Diffusion**: the movement of gas via a semipermeable membrane, from an area with a higher concentration to an area with lower concentration.

Ventilation

Ventilation is the delivery of oxygen and carbon dioxide in and out of the pulmonary airways. The effectiveness of breathing can be compromised by issues with the nervous, pulmonary, and musculoskeletal systems.

The Mechanics of Ventilation

Breathing is a result of the differences in atmospheric pressures and intrapulmonary pressures. Here is a summary of the mechanisms throughout process:

1. The intrapulmonary pressure is equal to the atmospheric pressure before inspiration - at approx. 760 mm Hg. Intrapleural pressure is 756 mm Hg.
2. Air is pulled into the lungs by the intrapulmonary atmospheric pressure gradient. The air is pulled into the lungs until both of the pressures.
3. Throughout inspiration, there is a contraction of the diaphragm and external intercostal muscles. This enlarges the thorax both vertically and horizontally, which causes the intrapleural pressure to decrease and the lungs to expand and fill the thoracic cavity.
4. Throughout regular expiration, the diaphragm gradually relaxes and the lungs and thorax return to their normal size and position. Throughout forced expiration, the contraction of the internal intercostal and abdominal muscles creates a reduction in thoracic volume. The compression of the lungs and thorax raises the intrapulmonary pressure to be greater than the atmospheric pressure.

The Influence of the Nervous System

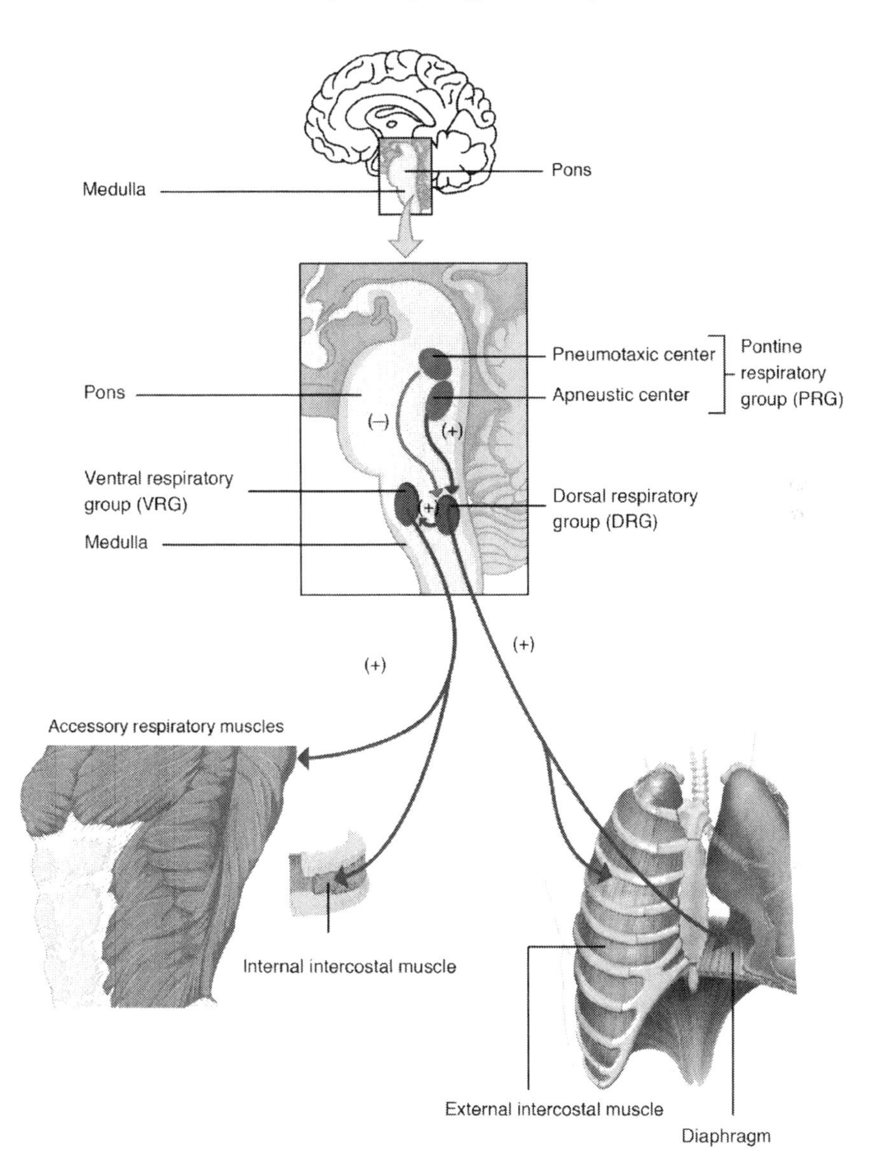

The process of involuntary breathing occurs due to stimulation that comes from the respiratory center in the medulla and pons in the brain. Carbon dioxide plays the main role in influencing breathing. The levels of carbon dioxide in the blood are monitored by the central chemical receptors in medulla. When there is an increase in the level of carbon dioxide, there is an increase in the rate and depth of breathing to remove the excess carbon dioxide.

The level of oxygen in the blood is monitored by the peripheral chemical receptors in the aorta and carotid arteries. When there is a decrease in the level of oxygen, there is an increase in the respiratory rate and depth, which elevates the levels of oxygen in the blood.

Pulmonary Influence

The distribution of airflow can be affected by a number of factors:

- The airflow pattern.
- The functional reserve capacity's volume and location.
- The magnitude of intrapulmonary resistance.
- Lung disease.

When the airflow is disrupted, the distribution of airflow will go down the path that has the least resistance.

Airflow Patterns

There are different patterns of airflows that affect the amount of airway resistance.

Laminar Flow

This is a linear pattern that happens at a low flow rate and gives a minimal amount of resistance. This type of flow occurs mostly in the in the peripheral airways of the bronchial tree.

Turbulent Flow

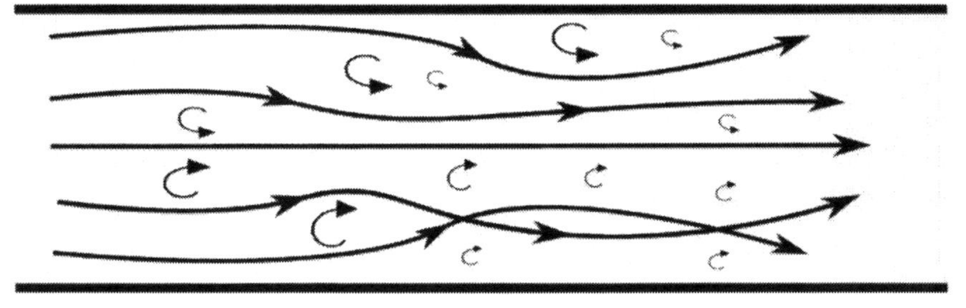

The turbulent flow pattern generates friction and elevates resistance. This flow is found in the trachea and large central bronchi. The turbulent flow may also occur in the smaller airways when they become constricted.

Transitional Flow

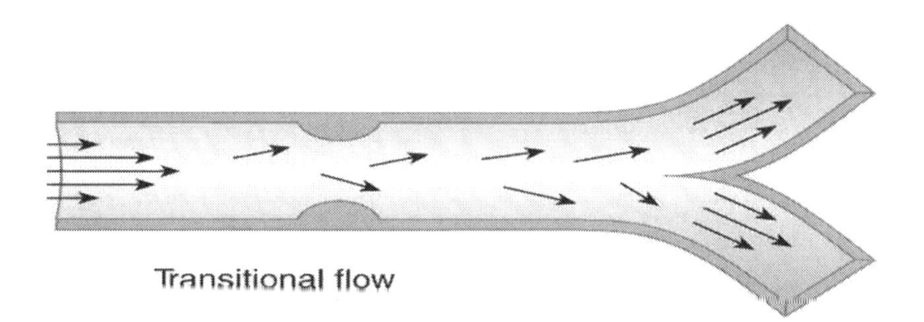

Transitional flow is a combined pattern, which occurs commonly in the larger airways at low flow rates, particularly where the airways meet or branch out.

Factors That Affect Airflow

Airflow can be affected by musculoskeletal and intrapulmonary factors that affect breathing. An example of this is in the condition called emphysema, which creates forced breathing. This activates the accessory muscles of respiration that require a greater amount of oxygen to function. The result of this is a decrease in ventilation efficiency and an increase in workload.

Respiratory muscle fatigue can be also be caused by other alterations in airflow. These include *changes in compliance,* in which there is an interference with the expansion of the lungs or thorax, and changes in resistance, in which there is interference with the airflow in the tracheobronchial tree. Both of these alterations can cause a reduction in tidal volume and alveolar ventilation.

Pulmonary Perfusion

Blood flow from the right side of the heart into the left side of the heart is known as pulmonary perfusion. The perfusion assists the external respiration. Alveolar gas exchange is made possible by normal pulmonary blood flow, however there are factors that can interfere with the transportation of gas to the alveoli. Examples of these factors are: a low cardiac output (less than 5 L/minute) and insufficient hemoglobin.

Ventilation-Perfusion Match

The transportation of oxygen and carbon dioxide can be positively affected by oxygen. Gravity results in a greater amount of unoxygenated blood travelling to the lower and middle lobes of the lung than to the upper lobes of the lung. This provides the explanation for why there are differences in ventilation and perfusion in different parts of the lungs. The areas in which perfusion and ventilation are alike are said to have ventilation-perfusion match. These areas have the most efficient gas exchange.

Diffusion

In the process of diffusion, both oxygen and carbon dioxide move between the alveoli and capillaries. This is movement that is from an area that has a higher concentration to an area of lower concentration. Oxygen moves through the alveolar and capillary membranes, and then dissolves in the plasmas, finally passing the red blood cell membrane. The movement of carbon dioxide is exactly the opposite.

Oxygen and carbon dioxide generally move through these layers easily. Oxygen travels into the bloodstream from the alveoli and it is taken up by hemoglobin in the red blood cells. The oxygen in the bloodstream displaces carbon dioxide. Carbon dioxide from the red blood cells then diffuses from the red blood cells, ending up in the alveoli.

The majority of oxygen binds with hemoglobin that forms oxyhemoglobin, but a small amount is dissolved in the plasma. The amount of oxygen that is dissolved in the plasma is measured using the *partial pressure of oxygen* in arterial blood (PaO_2).

Gas Exchange

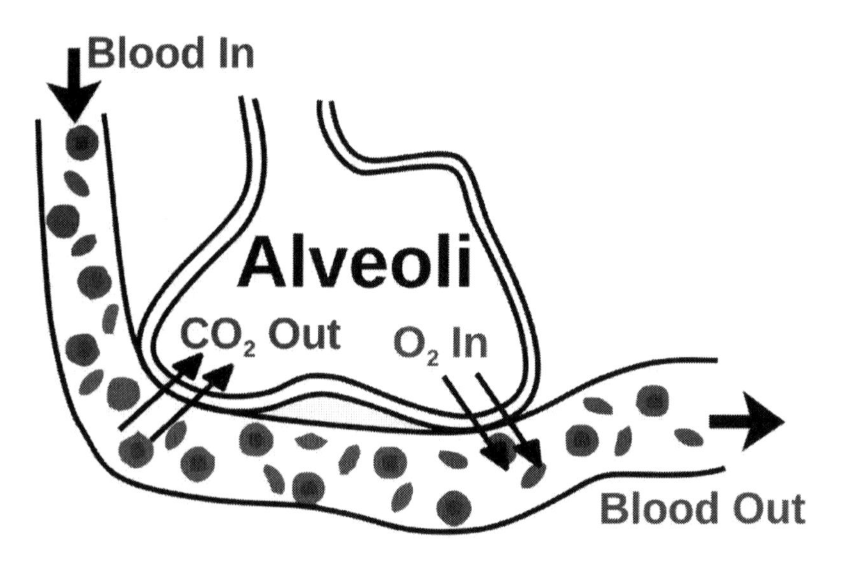

The exchange of gas takes place very quickly in the millions of thin-membrane alveoli that are in the respiratory units. Within the air sacs, the oxygen that is inhaled is diffused into the blood, and carbon dioxide is diffused out of the blood and exhaled into the air. The blood then travels around the body, circulating to delivery oxygen and pick up carbon dioxide. The final stage is the blood returning the lungs to be re-oxygenated.

Section 13: The Gastrointestinal System

The gastrointestinal (GI) system is made up of two main components: the GI tract (alimentary canal) and the accessory GI organs.

The GI Tract has two primary functions:

- **Digestion:** this is the process of breaking down food into simple chemicals that can be moved throughout the body when they are absorbed into the bloodstream.
- **Elimination:** this is the excretion of waste products through stool.

Alimentary Canal

The alimentary canal starts in the mouth and extends to the anus. It is a muscular tube that is hollow. Within the alimentary canal are the pharynx, esophagus, small intestine, stomach, and large intestine.

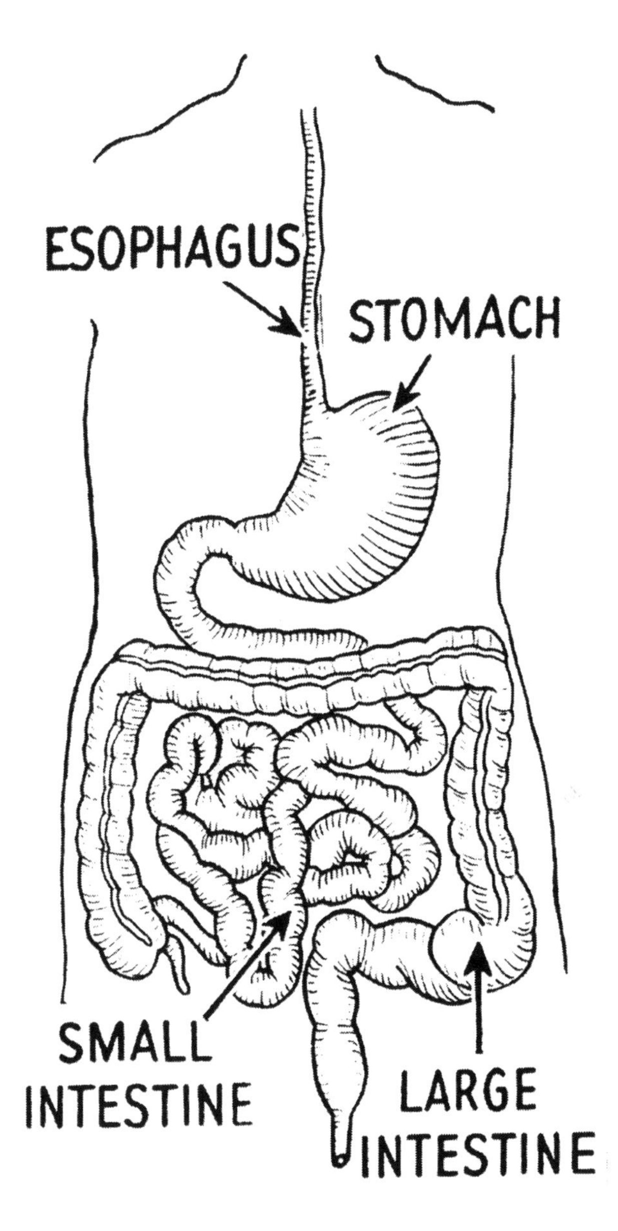

The complete GI system is made up of the alimentary canal, along with the accessory organs. The accessory organs are the liver, biliary duct system, and the pancreas.

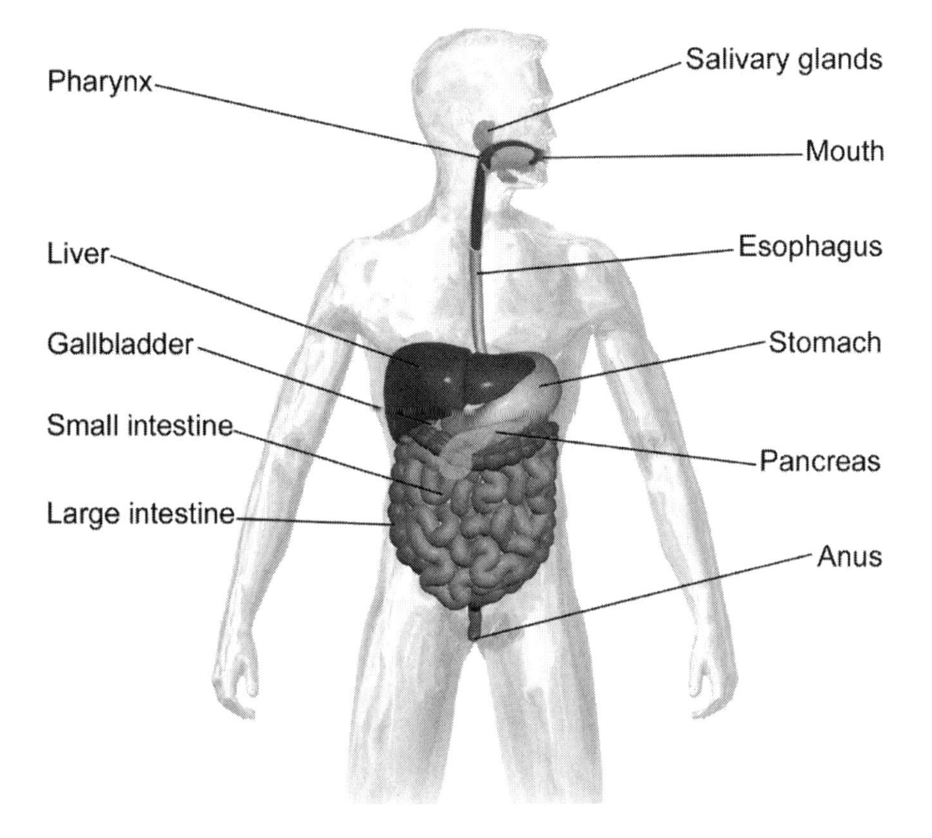
Pharynx
Salivary glands
Mouth
Liver
Esophagus
Gallbladder
Stomach
Small intestine
Pancreas
Large intestine
Anus

The Mouth

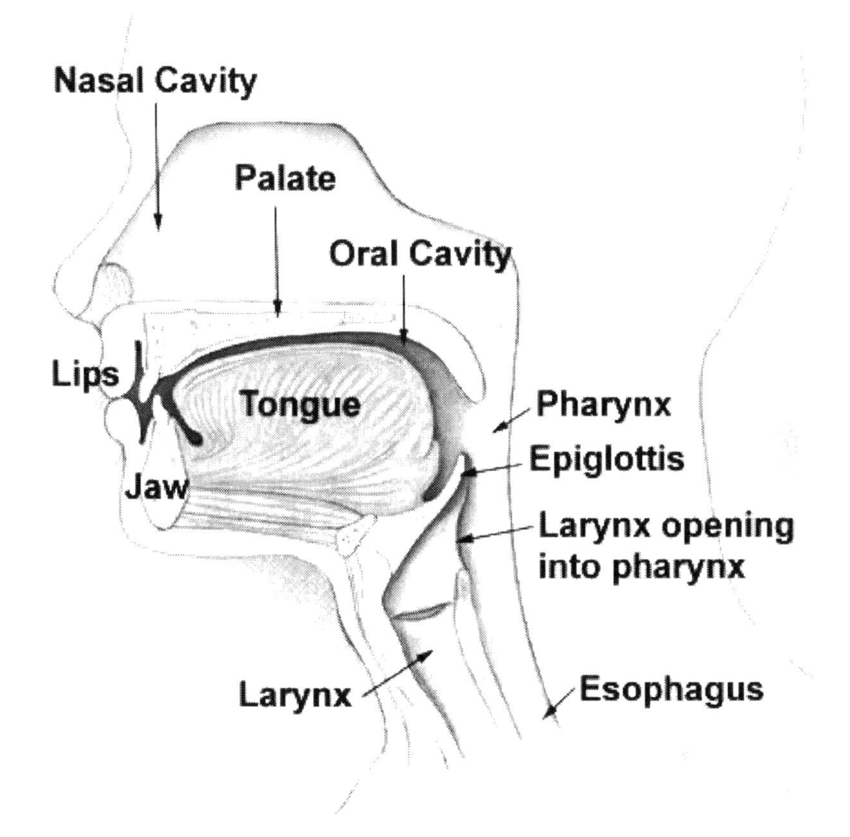

The mouth, also called the *oral cavity* or *buccal cavity*, is surrounded by the lips, root of the mouth (palate), cheeks, tongue, and teeth. The mouth is connected with the three main pairs of salivary glands (submandibular, parotid, and sublingual) by ducts. The salivary glands produce saliva that moistens food whilst chewing, and the mouth stimulates the breakdown of food.

Pharynx

The pharynx is a cavity that spans from the base of the skull to the esophagus. By taking food and pushing it toward the esophagus, the pharynx assists the swallowing process. When food arrives in the pharynx, there is a flap of connective tissue that is called the *epiglottis* that prevents aspiration by closing over the trachea.

Esophagus

The esophagus, a muscular tube, spans from the pharynx and through the mediastinum into the stomach. The passage of food from the pharynx into the esophagus is triggered by swallowing. In order for food to enter the esophagus, the *cricopharyngeal* sphincter must be relaxed. Liquids and solids are propelled into the esophagus and stomach by *peristalsis*.

Stomach

The stomach is a structure in upper left portion of the abdominal cavity, which is just beneath the abdominal cavity. It has a pouchlike collapsible structure, and its upper border is attached to lower section of the esophagus. The *greater curvature* is the stomach's lateral surface, and the *lesser curvature* is the stomach's medial surface. The size of the stomach can change as a result of *distention*.

The stomach is made up of four main areas:

- **The Cardia:** this is positioned in the area where the stomach and esophagus meet.
- **The Fundus:** this is an enlarged area to the left and above the stomach's esophageal opening.
- **The Body:** this is the middle area of the stomach.
- **The Pylorus:** this is the lower area of the stomach, which is positioned near where the stomach and duodenum meet.

The stomach serves a number of functions:

- The area where food is stored temporarily.
- The area where digestion begins.

- The breakdown of food into *chyme,* which is a semifluid substance.
- The moving of gastric contents to the small intestine.

Small Intestine

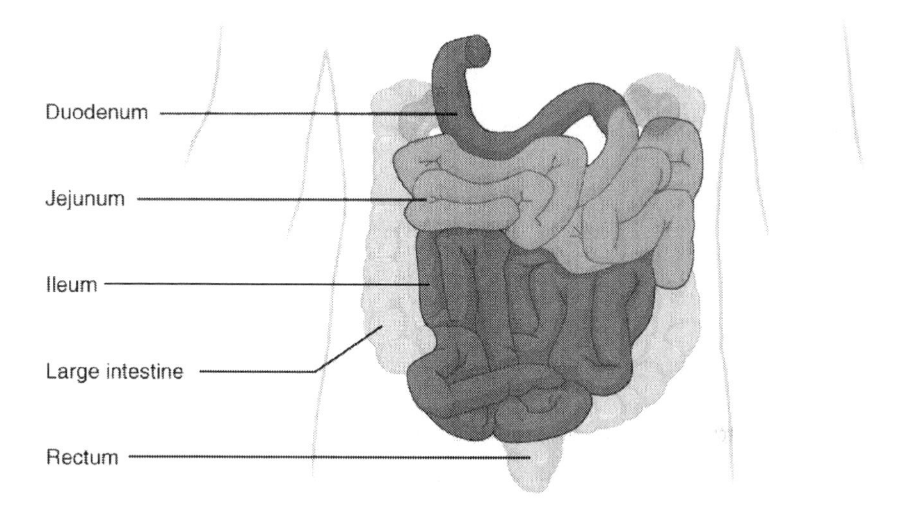

The small intestine is a tube that has a length of approx. 6 meters (20 inches). It has three main sections/divisions:

- **The Duodenum:** the division that is shortest.
- **The Jejunum:** the middle portion.
- **The Ileum:** the division that is the longest.

The intestinal walls is equipped with factors that increases the absorptive surface area:

- **Plicae Circulares:** these are circular folds of the intestinal mucosa/mucous membrane lining.

- **Villi:** these are finger-like projections located on the mucosa.
- **Microvilli:** these are micro cytoplasmic projects on the epithelial cells.

There are another three structures in the small intestine:

- **The Intestinal Crypts:** these are glands located in grooves that separate the villi.
- **The Peyer's Patches:** these are gatherings of lymphatic tissue located within the submucosa.
- **The Brunner's Glands:** these secrete mucus.

The Functions of the Small Intestine

- To complete the digestion of food.
- To absorb molecules of food through its wall into the circulatory system, and deliver the food molecules to the cells of the body.
- To secrete hormones that assist in the moderation of bile secretion, pancreatic fluid, and intestinal fluid.

Large Intestine

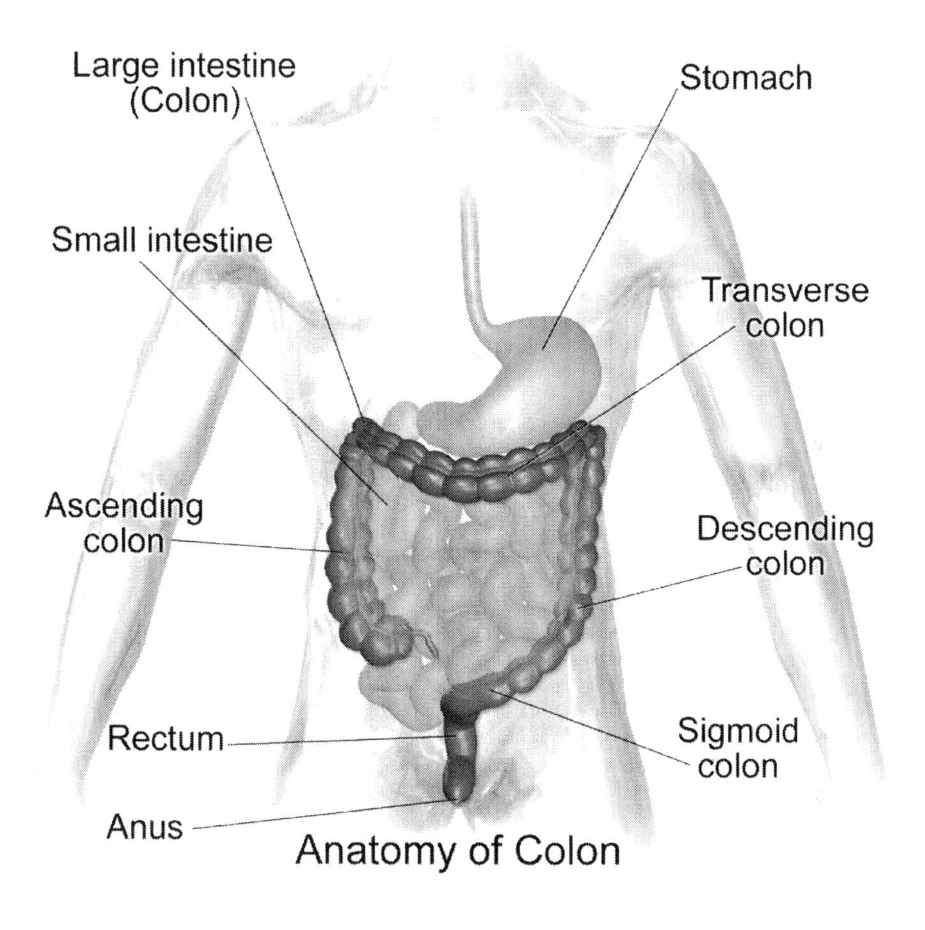

The large intestine spans from the valve near the ileum of the small intestine (ileocecal valve) to the anus. It is made up of distinct segments:

- **The Cecum:** this makes up the first few inches of the intestine and is a saclike structure.
- **The Ascending Colon:** this elevates on the right posterior abdominal wall. At the hepatic flexure it turns under the liver.
- **The Transverse Colon:** this is positioned over the small intestine, and it passes across the abdomen horizontally, underneath the liver, stomach, and spleen. It turns downward at the colic flexure.

- **The Descending Colon:** this begins near the spleen and it spans down the left side of the abdomen and into the pelvic cavity.
- **The Sigmoid Colon:** this descends down through the pelvic cavity and becomes the rectum.
- **The Rectum:** this represents the final few inches of the intestine and it ends at the anus. The anus is the large intestine's external opening for expulsion of waste products.

The large intestine serves the functions of mucus secretion, water absorption, and the elimination of digestive wastes.

The Wall Structures of the GI

A number of layers make up the wall of the GI tract: the mucosa, submucosa, visceral peritoneum, and the tunica muscularis.

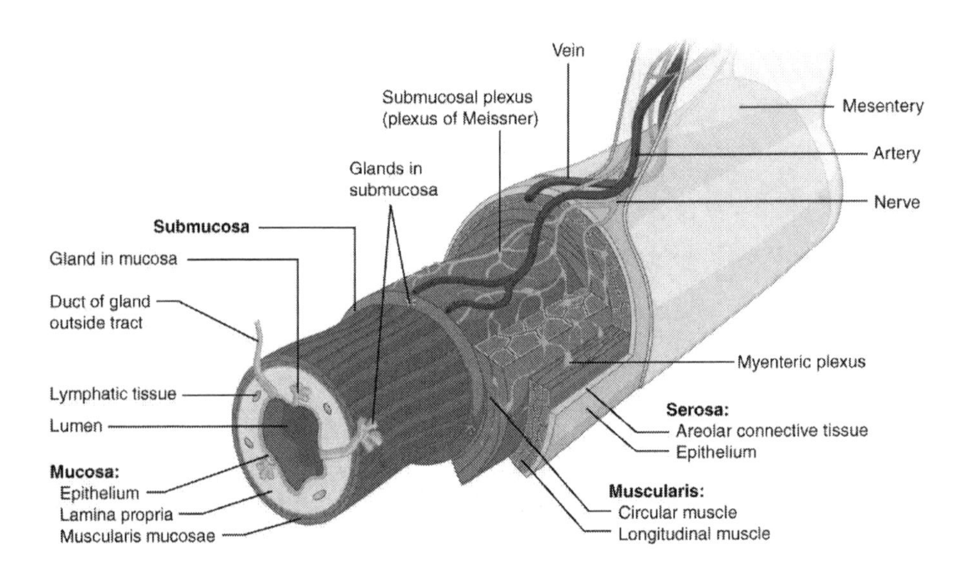

Mucosa

The mucosa is made up of epithelial and surface cells, along with loose connective tissue. *Villi* secrete gastric and protective juices, and absorb nutrients.

Submucosa

The submucosa surrounds the mucosa, and it is made up of loose connective tissue, lymphatic and blood vessels, and the submucosal plexus (a nerve network).

Tunica Muscularis

The tunica muscularis surrounds the submucosa and is made up of skeletal muscle in the pharynx, mouth and upper esophagus. In other areas of the tract, the tunica muscularis consists of muscle fibers (longitudinal and circular smooth).

The lumen length is shortened by fibers and the lumen diameter is decreased by circular fibers. At some points in the tract, sphincters are formed by the circular fibers that thicken.

The GI tract also has an outer covering that is called the *visceral peritoneum,* which covers the majority of the abdominal organs. It is positioned next to the *parietal peritoneum,* a layer that lines the abdominal cavity. In the esophagus and rectum, the visceral peritoneum is called the *tunica adventitia.* In other areas in the GI tract, it is called the *tunica serosa.* Around the blood vessels, lymphatics, and nerves, the visceral peritoneum becomes a double-layered fold. To prevent twisting, it attaches to the jejunum and ileum with the abdominal wall.

Parasympathetic Stimulation

Gut and sphincter tone is increased by parasympathetic stimulation of the vagus nerve and sacral spinal nerves. Along with this, parasympathetic stimulation increases the strength, frequency, and velocity of contractions in smooth muscle. It also increases secretory and motor activities.

Sympathetic Stimulation

Sympathetic stimulation causes a reduction in peristalsis and constrains the GI activity.

Accessory Digestive Organs

The liver, gallbladder, and pancreas are all accessory digestive organs that provide hormones, enzymes, and bile – these are all essential for digestion.

The Liver

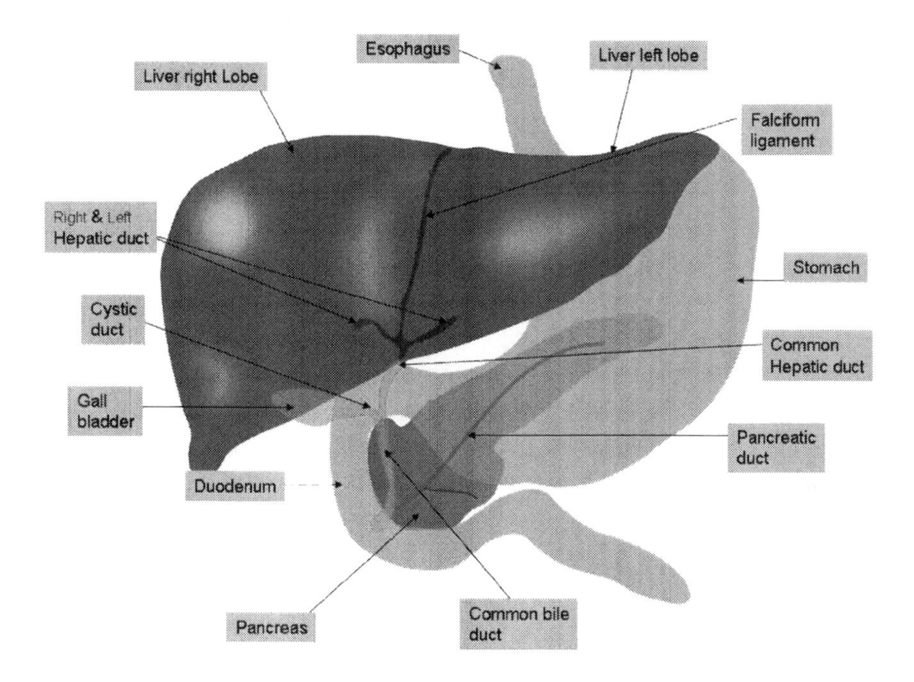

The liver is the largest gland in the body that weighs around 3 lbs. It is highly vascular and is surrounded by a fibrous capsule, located in the abdomen's upper right quadrant. The *lesser omentum* surrounds most of the liver and it anchors it to the stomach's lesser curvature. Passing through the lesser omentum is the hepatic artery, hepactic portal vein, common bile duct, and hepatic veins.

The liver is made up of four lobes.

- The left lobe
- The right lobe
- The caudate lobe, which is behind the right lobe.
- The quadrate lobe, which is behind the left lobe.

The Function of the Liver

The *lobule* is the functional unit of the liver. It is made up of a plate of hepatic cells (hepatocytes), which encircle around a central vein and spread outwards. The liver's capillary system is made up of *sinusoids,* which separate the hepatocyte plates from each other. Lining the sinusoids are *reticuloendothelial macrophages* (Kupffer cells), which remove bacteria and toxins in the blood that have entered through the intestinal capillaries.

The sinusoids provide transport for oxygenated blood, carrying them from the hepatic artery and to the portal vein. Blood that is unoxygenated exits through the central vein, and then flows to the inferior vena cava via the hepatic veins.

Below is an illustration of the liver lobules:

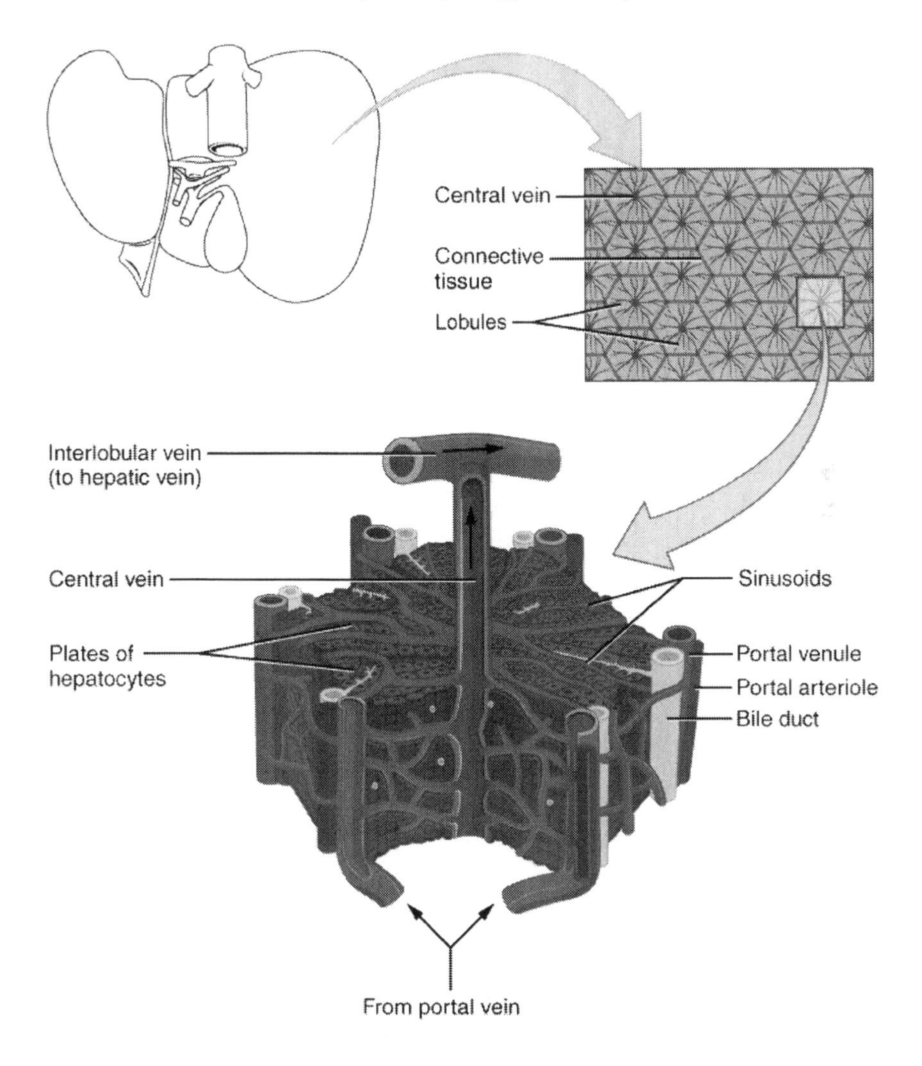

The Ducts

The ducts are a transport system that move bile through the GI tract. Bile is a liquid that has a green-like color – it is made up of water, cholesterol, phospholipids, and bile salts. Common hepatic ducts are formed as a result of bile ducts merging into the right and left hepatic ducts. The common hepatic duct meets the cystic duct from the gallbladder, which forms the common bile ducts, leading to the duodenum.

The Key Functions of the Liver

- Assisting the metabolism of carbohydrates.
- Detoxifying endogenous and exogenous toxins within plasma.
- Synthesizing plasma proteins, vitamin A, and nonessential amino acids.
- Providing storage for vitamin K, vitamin D, vitamin B12, and iron.
- Removing ammonia from body fluids and converting it into urea, which can be excreted through urine.
- Regulating levels of blood glucose.
- Secreting bile.

The Key Functions of Bile

- Breaking down (emulsifying) fat.
- Stimulating the intestinal absorption of cholesterol, fatty acids, and other lipids.

About 80% of bile salts are recycled into bile by the liver. These are combined with bile pigments and cholesterol, and this alkaline bile is continuously secreted by the liver. The production of bile may be increased as a result of stimulation from the vagus nerve, increased blood flow in the liver, the presence hormone secretin, and fat in the intestine.

The Gallbladder

The gallbladder is an organ that is joined to the liver's ventral surface by the cystic duct, and it is covered with visceral peritoneum. Bile produced by the liver is stored and concentrated by the gallbladder. The gallbladder also releases bile into the common bile duct so it can be transported to the duodenum.

GI Hormones

The GI structures secrete four hormones when stimulated, and each hormone plays a different role in the process of digestion. Below is a summary for each hormone:

GI Hormones

Hormone	Secreted By	Source & Stimulus	Target Organ	Respone
Gastrin	Stomach mucosa	Stomach in response to food	Stomach, small intestine	*release of HCL *Increase of intestine movement *release of pepsinogen
Secretin	Small Intestine	Duodenum in response to acidic chyme	Pancreas	*secrection of alkaline *digestive pro-enzyme *Inhibits intestine motility
Cholecystokinin (CCK)	Small Intestine	Intestinal cells in response to food	Pancreas, gallbladder	*Secretion of proenzymes and bile
Gastric Inhibitory Peptide (GIP)		Intestinal cells in response to fat	Stomach, Pancreas	*Insulin secretion *Inhibits gastric secretion and motality

Pancreas

The pancreas is an organ that has a fairly flat shape and it is located behind the stomach. The head and the neck of the pancreas span into the curve of the *duodenum* and the tail of the pancreas is positioned against the spleen. The pancreas serves both exocrine and endocrine functions.

- **Exocrene Function:** the exocrine function of the pancreas includes scattered cells that secrete over 1,000 ml of digestive enzymes each day. Lobes of the clusters (acini) and lobules of the cells that produce enzymes release secretions into the ducts that merge into the pancreatic duct.

- **Endocrine Function:** the endocrine function is served by the islets of Langerhans. These are positioned between the acinar cells.

An Illustration of the Biliary Tract

The combination of the gallbladder and the pancreas make up the biliary tract – the structures of this are shown below:

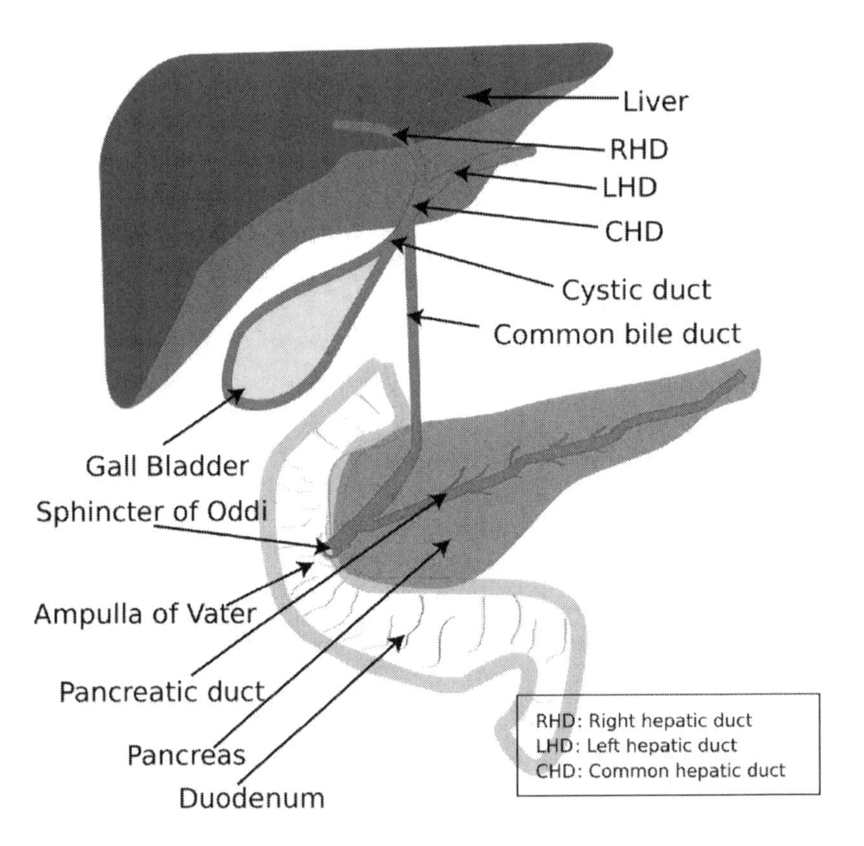

Islet Cells

The two types of islet cells are *alpha* and *beta*. These two types of cells are housed by over 1 million islets.

- **Beta Cells:** promote carbohydrate metabolism by secreting insulin.
- **Alpha Cells:** stimulate glycogenolysis in the liver by secreting glucagon.

The release of both of these hormones are stimulated by blood glucose levels, and they flow directly into the bloodstream

Pancreatic Duct

The pancreatic duct spans the entire length of the pancreas and it joins the bile duct from the gallbladder, prior to entering the duodenum. The rate and volume of pancreatic secretion is controlled by the release of the hormones secretin and cholecystokinin.

Digestion

The digestion process begins in the oral cavity, in which chewing (*mastication*), salivation (the start of start digestion), and swallowing (deglutition) happens. The hypopharyngeal sphincter and the upper esophagus relax when a person swallows. This allows food to enter the esophagus.

In the esophagus, peristalsis is activated by the glossopharyngeal nerve. This transports food down into the stomach. As the food travels through the esophagus, mucus is secreted by the glands in the esophageal mucosal layer. This provides lubrication for the bolus, along with protection for the mucosal membrane, which can result from food that has not been chewed completely.

The Cephalic Phase of Digestion

The cephalic phase of digestion begins as the food bolus travels toward the stomach. The stomach secretes hydrochloric acid and pepsin, which are digestive juices.

The Gastric Phase of Digestion

The stomach wall stretches as food enters it through the cardiac sphincter. This is the gastric phase of digestion, in which the stomach releases gastrin.

Gastrin:

Gastric activates the motor functions of the stomach, along with the secretion of gastric juice from the gastric glands. These digestive secretions are very acidic and are made up of mostly pepsin, hydrochloric acid, intrinsic factor, and proteolytic enzymes.

The Swallowing Process

Prior to peristalsis, the swallowing process must be initiated by the neutral pattern. Process is summarized as follows:

- Swallowing receptor areas are stimulated by food that is pushed to the back of the mouth. The receptor areas surround the pharyngeal opening.
- Impulses are then transmitted to the brain by the receptor areas via sensory areas of the trigeminal and glossopharyngeal nerves.
- The swallowing center of brain then transmits motor impulses to the esophagus via the trigeminal, glossopharyngeal, vagus, and hypoglossal nerves. This causes the swallowing to occur.

The Intestinal Phase of Digestion

Generally, a small amount of food absorption occurs in the stomach. Food is churned into small particles and mixed with gastric juices. This forms a substance called *chyme*. Following this, peristaltic waves transport the chyme into the antrum. This backs up against the pyloric sphincter prior to getting released into the duodenum. This stimulates the intestinal phase of digestion.

The Mechanisms of Gastric Secretion

The body of the stomach is positioned between the lower esophageal/cardiac sphincter and the pyloric sphincter. The fundus, body, antrum and body lie between these sphincters. To assist the stomach in carrying out its tasks, these areas have a number of rich mucosal cells.

2 to 3 liters of gastric juice is secreted daily by the cardiac glands, gastric glands, and the pyloric glands.

- Thin mucus is secreted by the cardiac gland and the pyloric gland.
- Hydrochloric acid, pepsinogen, mucus, and intrinsic factor is secreted by the gastric gland.

Specialized cells provide a lining for the gastric gland, gastric pits, and surface epithelium. Mucous cells located in the necks of the gastric glands generate thin mucus. Mucous cells located in the epithelium's surface generate an alkaline mucus. Both of these substances provide lubrication for food, along with protection for the stomach from corrosive enzymes.

There are also other secretions that occur. Gastrin is produced by *argentaffin*, which activates gastric secretion and motility. Pepsinogen is produced by chief cells, breaking down proteins into polypeptides. Hydrochloric acid and intrinsic factor is secreted by large parietal cells. Hydrochloric acid breaks down pepsinogen and prevents excess bacteria growth. Intrinsic factor stimulates vitamin B12 absorption within the small intestine.

The Emptying of the Stomach

The degree of stomach emptying is dependent on a number of factors, which includes the release of gastrin, the generation of neural signals when the wall of the stomach distends, and the *enterogastric reflex.* Throughout this process, secretin and gastric-inhibiting peptide is released by the duodenum. All of these function to reduce gastric motility.

The Role of the Small Intestine

Most of the digestion and absorption process is served by the small intestine. Carbohydrates, proteins, and fats are broken down by intestinal contractions and digestive secretions in the small intestine. This allows for these nutrients (with water and electrolytes) to be absorbed into the bloodstream, and then available for use by the body.

The Role of the Large Intestine

The food bolus starts its route through the large intestine where the cecum and ileum meet with the ileocecal pouch. Following this, the food bolus moves up the ascending colon and across the right abdominal cavity to the lower border of the liver. It moves across underneath the liver and stomach via the transverse colon. It then travels down the left abdominal cavity to the iliac fossa via the descending colon.

The large intestine does not produce any digestive enzymes or hormones. It continues the process of absorption via the blood and lymph vessels in the submucosa, where it absorbs remaining water in the colon, leaving 100 ml unabsorbed.

The mucosa within the large intestine also generates alkaline secretions, which lubricates the walls of the intestines as food is being pushed through. This protects the mucosa from acidic bacterial action.

Following this, the food bolus moves through the sigmoid colon into the abdominal cavity's lower midline. It then moves into the rectum and finalizes its journey at the anal canal. Through two sphincters, the anus opens to the exterior.

- **The Internal Sphincter:** this consists thick and circular smooth muscle that is under autonomic control.
- **The External Sphincter:** this consists skeletal muscle that is under voluntary control.

Bacterial Action

A number of bacteria are found in the large intestine: *Escherichia coli, Clostridium perfringens, Lactobacillus bifidus,* and *Enterobacter aerogenes.* These collectively assist in synthesizing vitamin K, along with breaking down cellulose in a carbohydrate that can be used. Bacteria also generate *flatus,* a gas that helps push stool toward the rectum.

Section 14: The Urinary System

The urinary system, also called the renal system, includes the following structures: the kidneys, ureters, bladder and urethra. The purpose of the urinary system is to remove excess fluid, waste products, and other substances from the body. Besides filtering substances out from the body through producing and expelling urine, the urinary system is also responsible for performing endocrine functions and balancing water content, amongst other things.

Female urinary system:

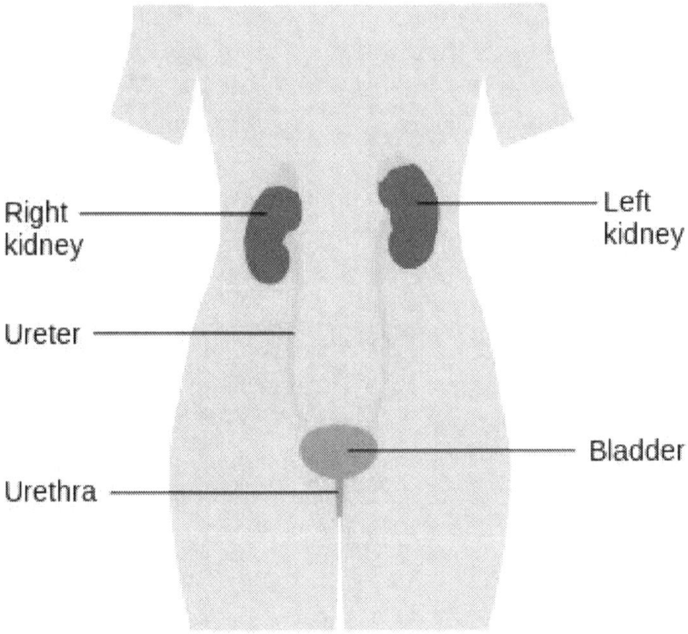

Right kidney
Left kidney
Ureter
Bladder
Urethra

Male urinary system:

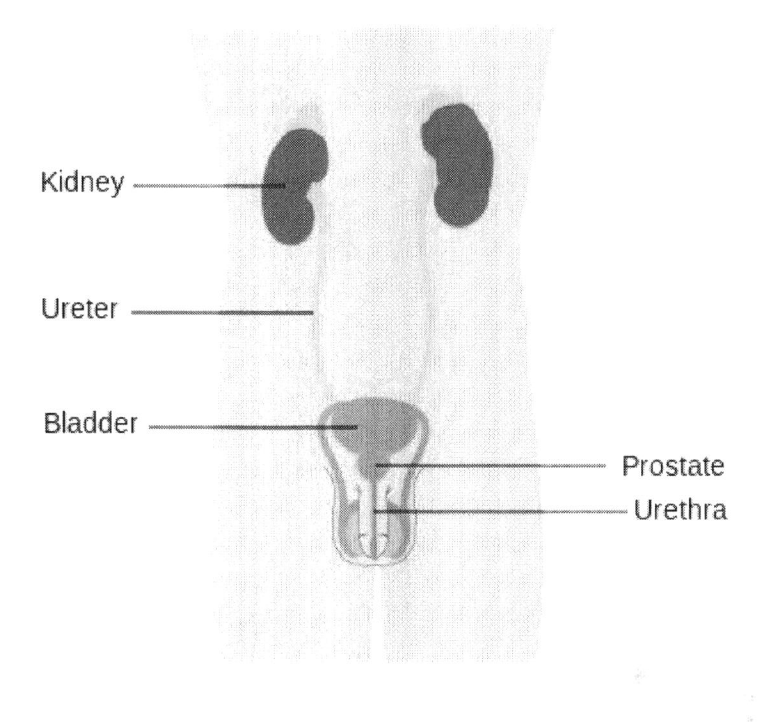

The Kidneys

The kidneys are organs located to the dorsal side of the abdominal cavity, on the right and left side of the abdomen. They are responsible for removing waste products from the blood in the form of urine, regulating the chemical composition of blood, producing the hormone erythropoietin, producing the enzyme renin, converting vitamin D into a more active compound, as well as maintaining the fluid, electrolyte and acid-base balances.

Kidneys are highly vascular, which means that they contain a great number of blood vessels. Blood supply to the kidneys is by way of the renal artery which subdivides into several branches.

Each kidney comprises three regions: the renal cortex, the renal medulla, and the renal pelvis.

- **The Renal Cortex:** The renal cortex is the outer layer of the kidney which lies between the renal capsule and the renal medulla. It contains blood-filtering mechanisms and is protected by layers of fat and a fibrous capsule. Beneath the renal cortex lies the renal medulla.
- **The Renal Medulla:** The renal medulla is the innermost part of the kidney which is split up into different sections called the **renal pyramids.** Renal pyramids are cone-shaped tissues of the kidney which empty into minor calyces. Two to three minor calyces come together to form one major calyx. Renal pyramids secrete urine. Within the urinary system, urine passes through the calyces' channel from the renal pyramids to the renal pelvis.
- **The Renal Pelvis:** The renal pelvis is the innermost layer of the kidney. Urine is discharged into the renal pelvis before being funneled into the ureter.

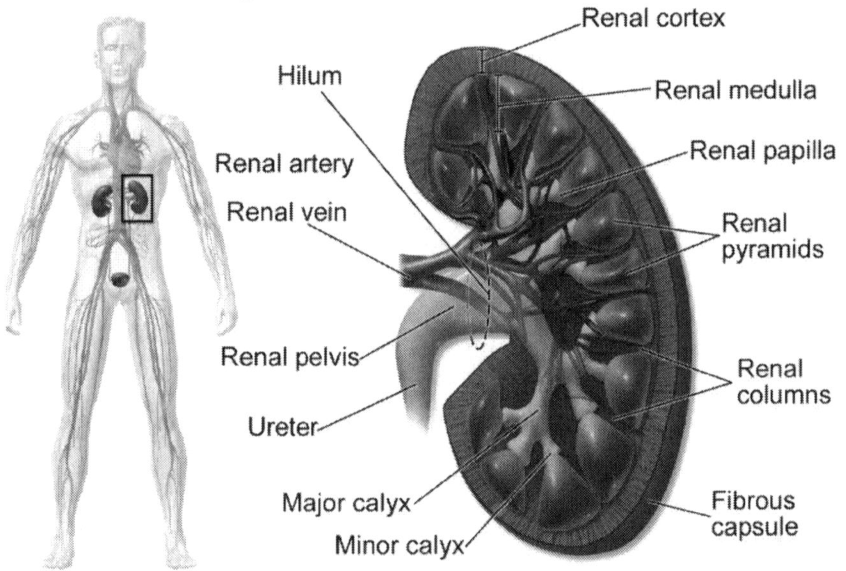

The adrenal glands sit on top of the kidneys. The adrenal glands affect the urinary system by influencing blood pressure and by regulating the retention of sodium and water by the kidneys.

The Nephron

The nephron is the basic structural and functional unit of the kidney and the site of urine formation. Nephrons perform two main functions: the secretion and reabsorption of ions and the filtration of fluids, electrolytes, acids, bases, and waste products into the tubular system. The process by which the kidneys filter the blood is called **glomerular filtration.**

Each nephron consists of a **glomerulus** and a **collecting tubule**. The glomerular filtrate passes through these tubules before emerging as urine. The glomerulus (the tubular apparatus) is found inside the **Bowman's capsule**, which is a capsule-shaped membranous structure.

Each nephron is divided into three portion:

- The **proximal convoluted tubule** is the nephron portion which is nearest to the glomerular capsule.
- The **loop of Hemle** is the second portion.
- The **distal convoluted tubule** is the third portion which is farthest away from the glomerular capsule. The distal convoluted tubule joins up with the far end of another nephron, forming a larger collecting tubule.

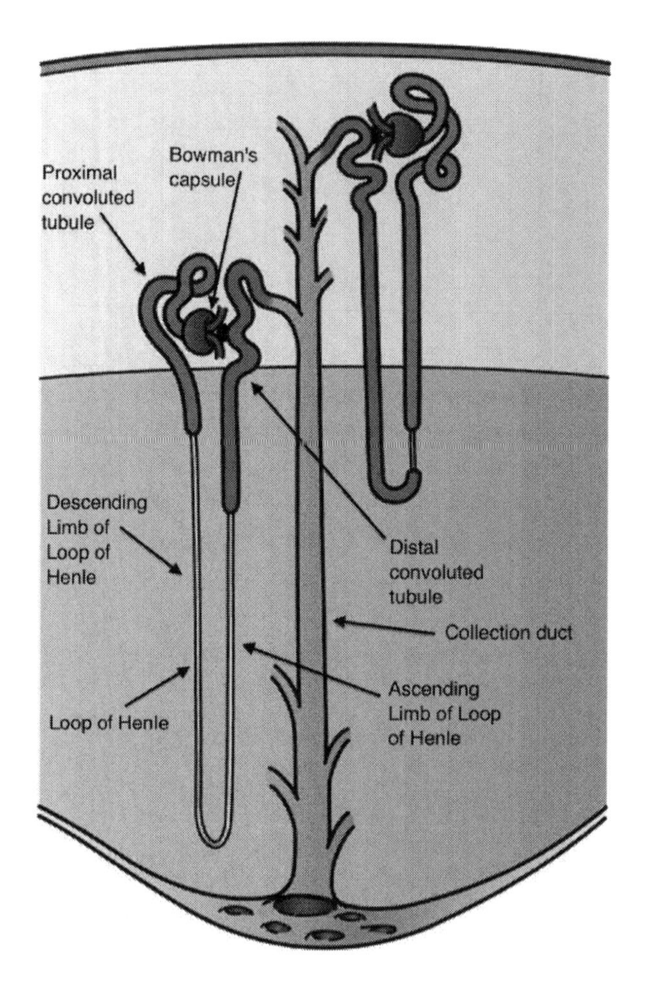

The glomerulus (located inside the Bowman's capsule), the proximal convoluted tubules and the distal convoluted tubules are all situated in the renal cortex. The loops of Henle are located in the renal medulla.

The Ureters

The ureters are fibro-muscular ducts which connect the kidney to the bladder. Urine passes from the kidneys through the ureter and into the bladder. Peristaltic waves (involuntary smooth muscle contractions) transport urine along the ureters and into the urinary bladder.

The Bladder

The bladder is a muscular organ located in the pelvis which stores urine. The bottom of the bladder forms a **trigone**, a triangular region, which contains three openings. Two of these openings connect the bladder to the ureters and the third connects the bladder to the urethra.

The Urethra

The urethra is the small duct by which urine passes from the bladder to the outside of the body. The urethra connects the bladder to the **urinary meatus**, that is, the external opening of the urethra. In females, the urethra is located inside the anterior wall of the vagina. In males, the urethra passes through the prostate gland from where it extends into the penis. The male urethra is a passageway for both urine and semen.

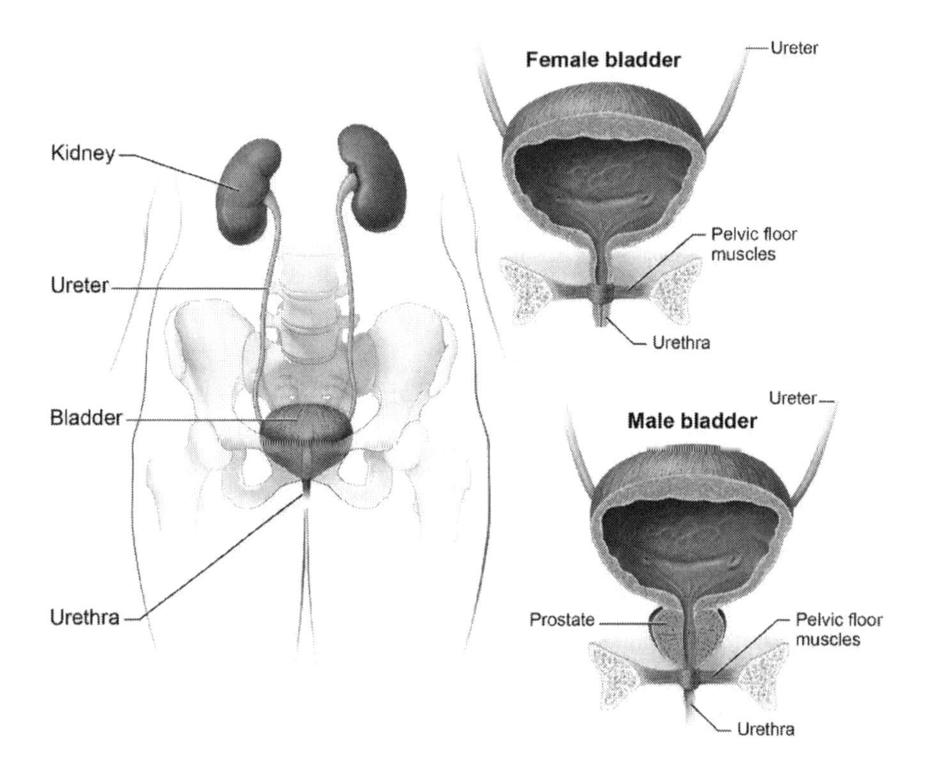

Three Processes of Urine Formation

Urine consists primarily of ammonium ions, bicarbonates, calcium, chloride, creatinine, magnesium, phosphates, sodium, sulfates, uric acid, urobilinogen, and water. Other substances may also enter urine, such as leukocytes and RBCs.

Urine is formed through three steps: **glomerular filtration, tubular reabsorption,** and **tubular secretion.**

1. **Filtration:** Glomerular filtration occurs as blood flows into the glomerulus. Sodium and glucose are reabsorbed through active transport from the proximal convoluted tubules. Reabsorption of water is caused by osmosis.

2. **Reabsorption:** During tubular reabsorption, substances from the glomerular filtrate move from the distal convoluted tubules into the peritubular capillaries. Sodium, glucose and potassium are reabsorbed through active transport. Antidiuretic hormone stimulates the reabsorption of water.
3. **Secretion:** During tubular secretion, the substances move from the peritubular capillaries into the tubular filtrate. The peritubular capillaries secrete ammonia and hydrogen which are then transported into the distal tubules by way of active transport.

Physiologic Mechanism of Urine Excretion

Nephrons remove waste products from the blood. The kidneys receive blood which contains waste from the renal artery. After passing through smaller blood vessels and the nephrons, the filtered blood is reabsorbed by the peritubular capillaries. **Peritubular capillaries** are tiny blood vessels located around the nephrons. The filtered blood then re-enters circulation through the renal vein which then empties into the inferior vena cava.

The kidneys excrete waste products that the nephrons remove from the blood. These waste products are combined with other waste fluids to form urine. The urine passes through the ureters and into the urinary bladder through a process called peristalsis. **Peristalsis** is the involuntary relaxation and constriction of muscles of the intestine and other canals.

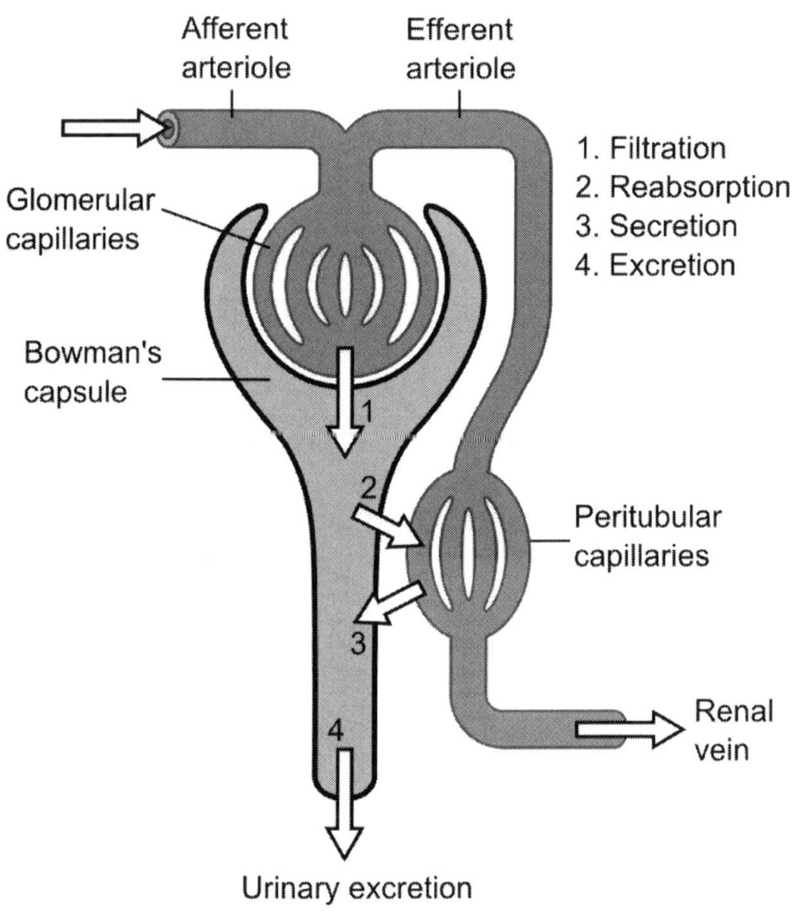

The micturition reflex:

Once the bladder has filled with urine, parasympathetic nerves cause the bladder to contract and the sphincter to relax. This relaxation in combination with a voluntary stimulus (the intentional and learned process of urination) allows urine to pass into the urethra for elimination from the body.

The Role of Hormones in the Urinary System

Hormones play a big role in the urinary system. The hormones affecting the urinary system include the following: antidiuretic hormone, aldosterone, erythropoietin, angiotensin I and angiotensin II.

- **Antidiuretic hormone (ADH):** ADH helps regulate the level of urine production by causing the kidneys to retain water. Low levels of ADH decrease water reabsorption and can lead to dilute urine, whereas high levels of ADH increases water reabsorption as well as urine concentration.
- **Aldosterone:** Aldosterone, which is secreted in the adrenal cortex, facilitates tubular reabsorption. It does this by regulating sodium retention and potassium secretion. When serum potassium levels are rising for example, the adrenal cortex will start to increase its production of aldosterone. This facilitates sodium retention which in turn raises blood pressure.
- **Erythropoietin:** When arterial oxygen tension is low, the kidneys respond by secreting erythropoietin. Erythropoietin travels to the marrow where it stimulates the production of RBC (also called erythrocytes). RBCs help increase the blood's oxygen-carrying capacity.
- **Angiotensin I and angiotensin II:** These hormones are part of the renin-angiotensin-aldosterone system.

The renin-angiotensin-aldosterone system.

The renin-angiotensin-aldosterone system helps regulate blood pressure, water and sodium levels in the body.

- Low blood pressure and low sodium levels in the kidneys are two main factors that cause the kidneys to secrete the enzyme renin.

- Renin, which enters blood circulation, promotes the production of angiotensin I which is then converted into angiotensin II as it circulates through the lungs.
- Angiotensin II is a powerful chemical which causes contraction of the muscles that surround blood vessels.
- This contraction results in the narrowing of the blood vessels which in turn raises blood pressure.
- Angiotensin II also stimulates the production of the hormone aldosterone in the adrenal cortex.
- Aldosterone helps regulate water and sodium balance in the body by stimulating the absorption of sodium by the kidneys.
- Increased water and sodium retention lowers sodium levels which in turn lowers blood pressure.

Low blood pressure and low sodium levels stimulates the kidneys to secrete renin, thus readjusting homeostasis.

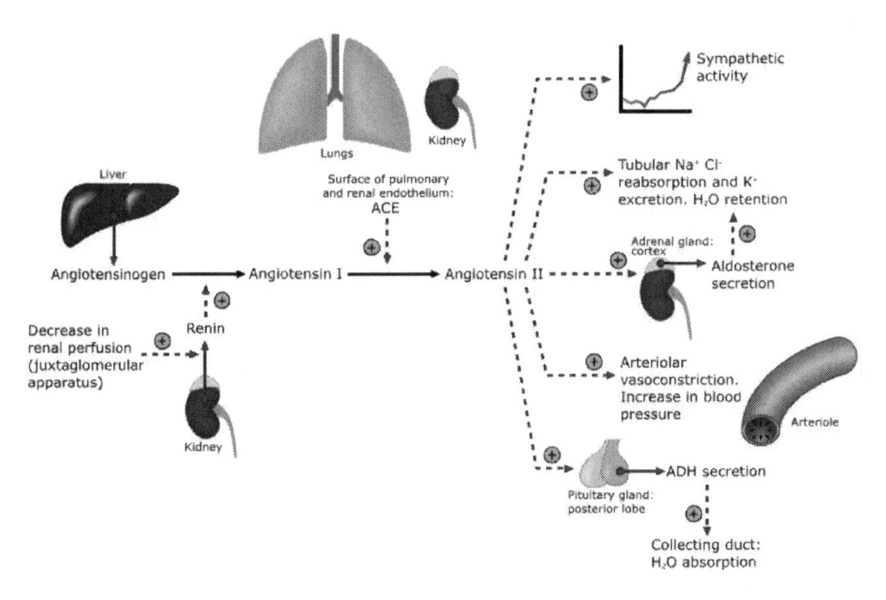

Section 15: The Reproductive System

The reproductive system consists of the organs and glands in the body responsible for reproduction. To begin with, this section will look at the structure and function of the female reproductive system before moving on to the male reproductive system.

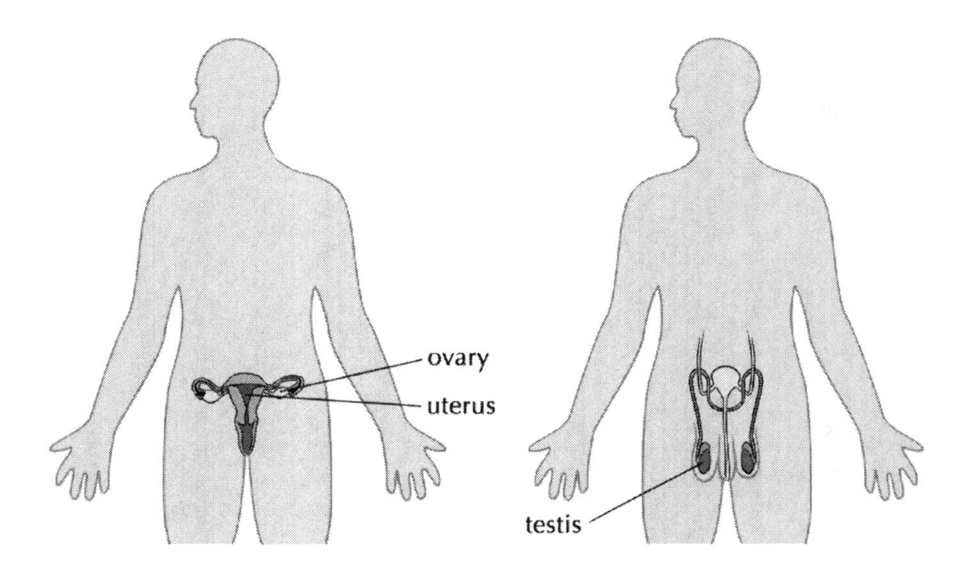

The Female Reproductive System

The female reproductive system is located largely inside the pelvic cavity. The consists of the vulva, vagina, cervix, uterus, fallopian tubes, and the ovaries. The vulva is the female external genitalia whereas the other structures make up female internal genitalia.

The Vulva

The vulva consists of the vaginal opening, mons pubis, clitoris, labia majora, labia minora, and adjacent glands.

- **Mons Pubis:** The mons pubis is the rounded mass of fatty connective tissue which covers the joint of the pubic bones.
- **Labia Majora:** The labia majora are the two larger outer folds of the vulva which consist of adipose and connective tissue.
- **Labia Minora:** The labia minora are the two moist inner folds of the vulva which are made up of mucosal tissue. The labia minora contains sebaceous glands which secrete sebum – oily matter which also acts as a bactericide. These glands are highly responsive and swell in response to sexual stimulation.
- **Clitoris:** The clitoris is a small organ which lies at the anterior end of the vulva. The fold of skin surrounding the clitoris is called **prepuce.**
- **Vestibule:** The vulval vestibule is the part of the vulva which lies between the labia minora and the vaginal opening. In the center of the vestibule is the **vaginal orifice** which may or may not be covered by a tissue membrane called the **hymen.**
- **Bartholin's glands:** Bartholin's glands are found slightly posterior and to the left and right of the vaginal opening. These glands secrete mucus which serves as lubricant to the vagina.
- **Urethral meatus:** The urethral meatus is the duct below the clitoris through which urine exits the body.

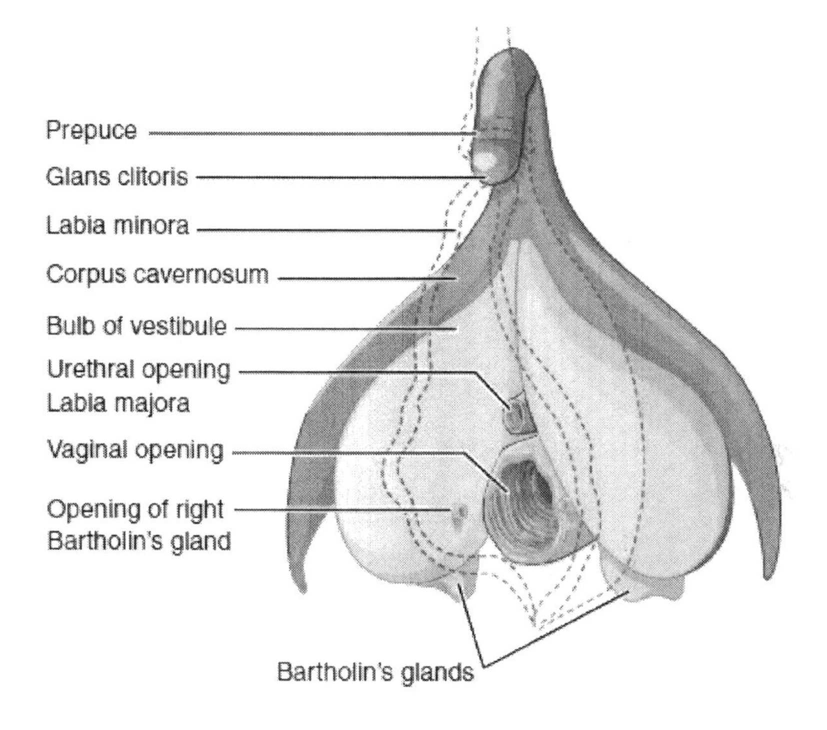

Vulva: Internal anteriolateral view

The Vagina

The vagina is the muscular tube which is located between the rectum and the urethra. The vagina leads from the female external genitalia to the cervix of the uterus. The wall of the vagina consists of three tissue layers: epithelial tissue, loose connective tissue and muscle tissue. The vagina has three main function: to receive the penis during sexual intercourse, to channel menstrual discharge from the uterus, and to serve as a birth canal during childbirth.

The Cervix

The cervix is a narrow cylinder-shaped passage that connects the vagina and the uterus. The lower cervical opening is called the **external os** and the upper cervical opening is called the **internal os.**

The Uterus

The uterus is a muscular organ located between the bladder and the rectum. The uterus contains three layers. The outer layer is a thin layer of epithelial cell tissue called the **serosa** or **perimetrium.** The middle layer is a muscular layer called the **myometrium,** which primarily consists of smooth muscle cells. The inner lining of the uterus is composed of a mucous membrane called **endometrium.**

The Ovaries

The ovaries, which are located on either side of the uterus, are the female reproductive organ in which eggs (ova) are produced. The size, position, and shape of these sex cell-producing organs vary with age.

The Fallopian tubes

The fallopian tubes, which are the site of fertilization, are the tubes through which ova travel from the ovaries to the uterus.

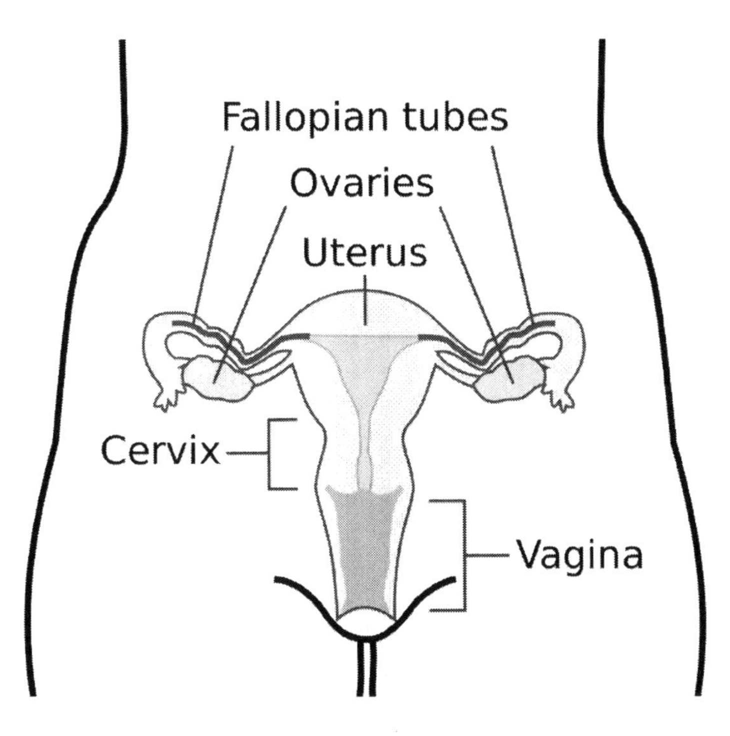

Mammary Glands

Mammary glands are milk-producing glands found inside female breasts. In the male, mammary glands are also present but are underdeveloped. Each breast consists of **fibrous, adipose**, and **glandular tissue**. A healthy female breast has glandular tissue which is made up of 12 – 20 sections called **lobes.** These lobes contain clustered **acini**, which are small saclike cavities that secrete milk.

Hormones and the Male Reproductive System

The female menstrual cycle is regulated by a complex interaction of hormones which are secreted by the hypothalamus, ovaries, and pituitary gland. The hormones involved in the the menstrual cycle include **estrogen, progesterone, FSH** and **LH.** In females, the

reproductive cycle lasts around 28 days and includes two distinct cycles: the **ovarian cycle** and the **uterine cycle**.

The Ovarian Cycle

The ovarian cycle is responsible for producing the hormones which then go on to control the uterine cycle.

- **Day 1 – 13:** In response to low estrogen levels, FSH stimulates the development of a follicle and LH. An increase in LH levels promotes the maturation of an oocyte in one of the ovaries. Once the follicle has reached a certain stage of development, it starts secreting estrogen. Once the estrogen level has reached a certain level, a negative feedback mechanism involving the hypothalamus slows the secretion of LH and FSH.
- **Day 14:** Once the follicle has fully developed, a spike in LH and FSH levels occurs. This causes the follicle to rupture, which releases the ovum and initiates ovulation.
- During ovulation, which typically takes place on day 14, the oocyte is released. After ovulation, the follicle from which the oocyte was released becomes a corpus luteum. The corpus luteum is a hormone-secreting structure which, unless pregnancy occurs, degenerates after a few days. The corpus luteum secretes progesterone.
- **Day 17:** Once progesterone levels have reached a certain threshold, usually around day 17, the hypothalamus stops the secretion of LH. At this point, the corpus luteum begins to degenerate.
- **Day 26:** Roughly on day 26, the corpus luteum has disappeared and estrogen and progesterone levels are at their lowest.
- **Day 29:** When the hypothalamus senses low estrogen levels, it secretes GnRH (gonadotropin-releasing hormone) which stimulates the anterior pituitary gland to secrete FSH which in turn stimulates the development of another follicle. The menstruation cycle starts over.

The Uterine Cycle

The function of the uterine cycle is to prepare the uterus for possible fertilization.

- **Day 1 - 5 (Menstruation):** The first five days of the uterine cycle is the period of menstruation. During this period, estrogen and progesterone levels are lowest. As the levels of sex hormones drop, cells and tissues undergo autolysis (self-destruction). As a result of this, the uterus wall breaks down and blood vessels rupture. This causes bleeding during a period. The menstrual flow passes through the cervix, and out of the body through the vagina.
- **Day 6 - 14 (Proliferative Phase)**: The phase of the menstrual cycle after menstruation is called the proliferative phase. During this phase, estrogen production peaks. The developed follicle secretes estrogen which rebuilds stimulates the endometrium (the lining of the uterus) to regenerate new tissue.
- **Day 15 - 28 (Secretory Phase)**: The phase after ovulation occurs is called the secretory (luteal) phase. During this phase, the corpus luteum secretes progesterone which further thickens the endometrium. During this phase, the uterus also secretes a thick mucus. If fertilization occurs, the thick endometrium and the mucus will capture the fertilized egg and assist in its proper implantation in the uterus. If fertilization doesn't occur, the corpus luteum degenerates and progesterone levels decline. This causes the endometrium to (re-)shed its functional layer, and the cycle starts again.

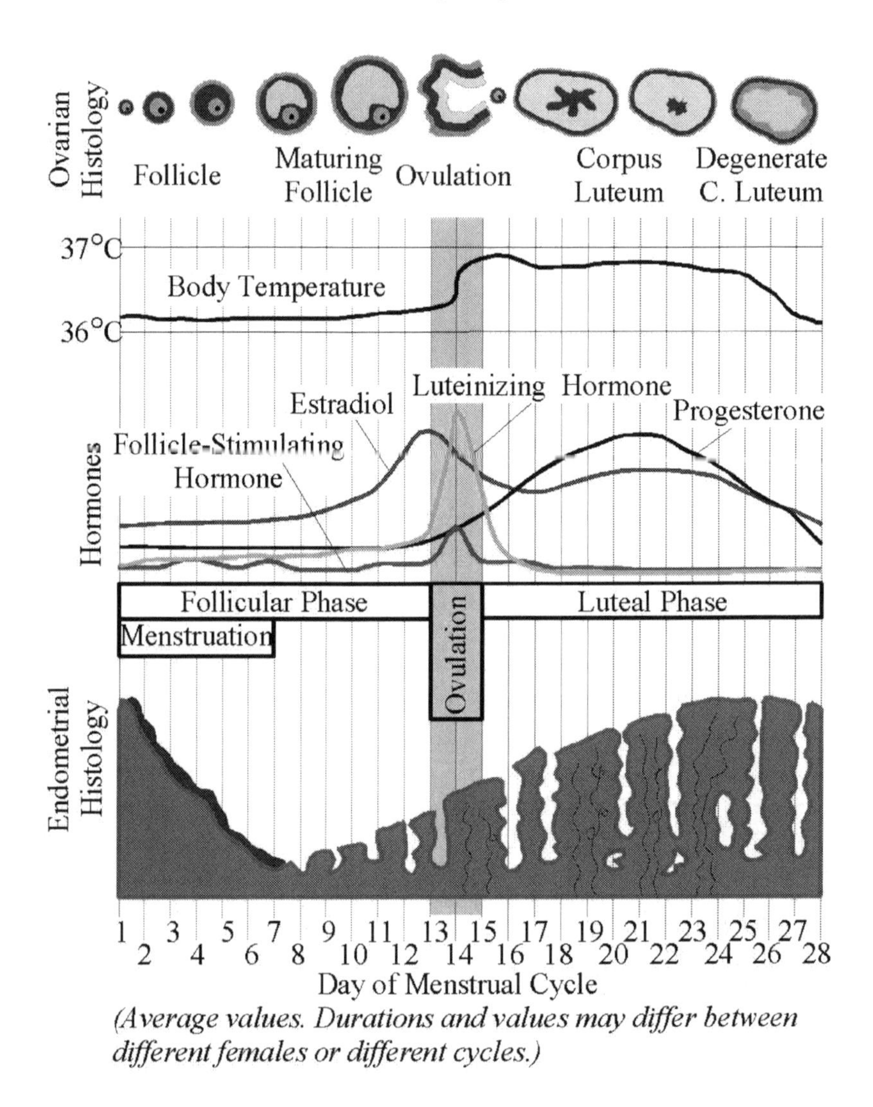

(Average values. Durations and values may differ between different females or different cycles.)

The Male Reproductive System

The male reproductive system consists of the organs and structures involved in the production, transfer, and introduction of sperm into the female reproductive system. It comprises the penis, scrotum, prostate gland, and inguinal structures.

The Penis

The penis is the intromittent organ which comprises a **penile shaft** and a **glans penis** (the tip).

- The penile shaft consists of three columns of **erectile tissue** which are held together by fibrous tissue. Two masses of erectile tissue called **corpora cavernosa** form the bulk of the penis. The third mass of erectile tissue which surrounds the urethra is called **corpus spongiosum.**
- The glans penis is the bulbous structure at the distal end of the penile shaft. The glans penis is formed from the corpus spongiosum and is highly sensitive to stimulation. The urethral meatus opens through the glans penis, allowing ejaculation and urination.

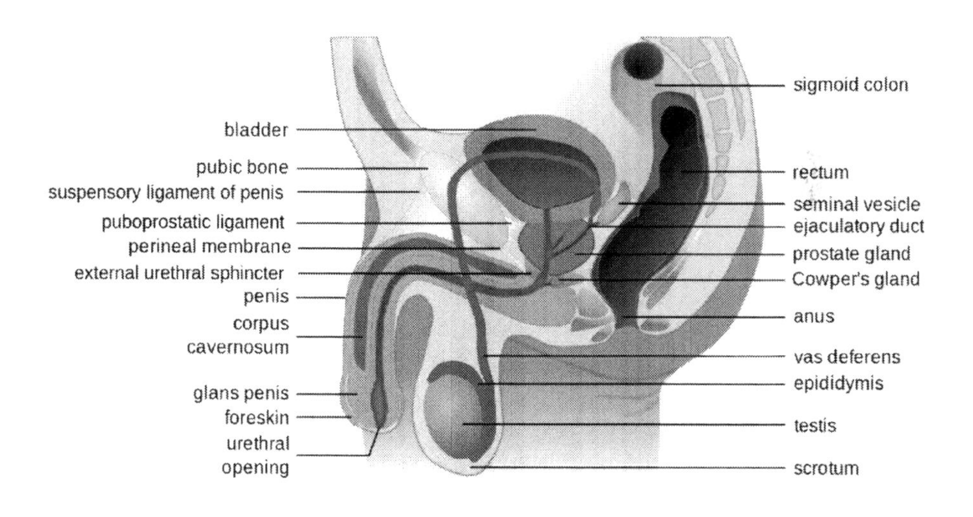

The Scrotum

The scrotum is the extra-abdominal pouch skin which contains the testicles. It is located posterior to the penis and anterior to the anus. An internal partitioning divides the scrotum into two chambers (sacs). Each sac contains a **testis**, an **epididymis** and a

spermatic cord.

The spermatic cord, a connective tissue sheath, consists of a bundle of nerves, blood vessels, and lymph vessels which connect the testicles to the abdominal cavity. The spermatic cord also contains the **vas deferens**, which is the duct that transports sperm from the testicle to the urethra.

The Testes

The testes are the organs which produce male reproductive cells called **spermatozoa**. The testes are covered by two layers of fibrous connective tissue. The outer layer is called **tunica vaginalis** and the inner layer is called **tunica albuginea.** Extensions of the tunica albuginea divide each testis into numerous lobules, each of which contains one to four convoluted tubes, called the **tubuli seminiferi.** The tubuli seminiferi, also called seminiferous tubules, are the site of spermatogenesis – the production of mature spermatozoa.

Cross Section of Testicle

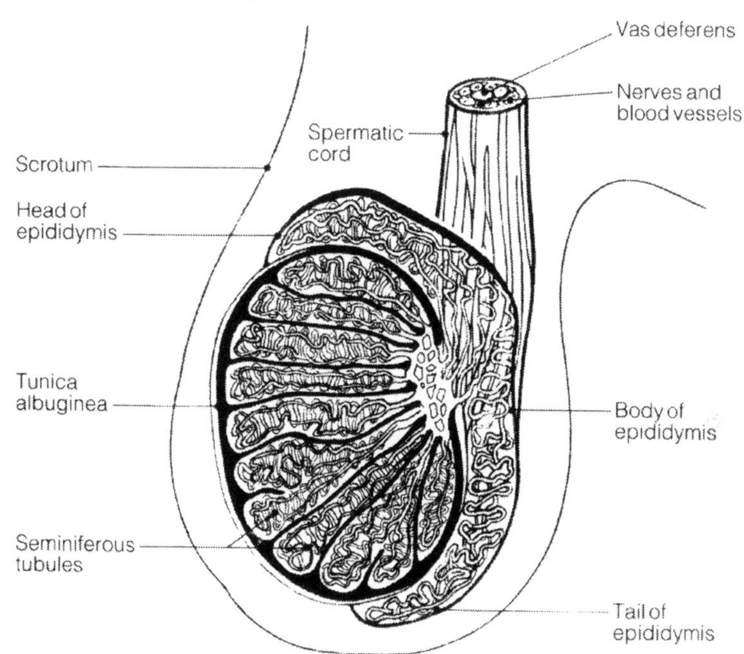

The Duct System

The duct system comprises the **epididymis** and the **vas deferens**. The epididymis, located behind the testis, is a convoluted duct through which sperm pass into the vas deferens. During ejaculation, contraction of the smooth muscle in the epididymis occurs, which ejects spermatozoa into the vas deferens. The epididymis is continuous with the vas deferens. The vas deferens is the duct that transports sperm from the epididymis of each testicle to the urethra.

Accessory Genital Ducts

Other internal organs of the male reproductive system, sometimes referred to as accessory organs, produce most of the seminal fluids. They include the seminal vesicles, bulbourethral glands, and the prostate gland. These three glands secrete fluids that form semen.

- **Seminal Vesicles:** The seminal vesicles are located on each side of the male urinary bladder, at the juncture of the bladder and the vas deferens. Seminal vesicles produce around 60% of the fluid component of semen. The remaining 30% is produced by the prostate gland.
- **Prostate Gland:** The prostate gland, which lies at the base of the bladder, is an organ which surrounds the urethra in males. It secretes alkaline fluid called **prostatic fluid** that forms part of semen and also helps control urination.
- **Bulbourethral Glands:** Also called Cowper's glands, bulbourethral glands are located behind and to the side of the urethra, near the bulb of the penis. Bulbourethral glands secrete viscid fluid which also forms part of semen.

Spermatogenesis

Spermatogenesis is the process of sperm formation. Spermatogenesis begins during puberty and typically continues throughout life. The quantity and quality of the sperm however, varies with health and age. The process of spermatogenesis is divided into four stages.

1. During the first stage, **spermatogonia** (germinal epithelial cells) grow and develop into (primary) **spermatocytes** (sperm cells). Both spermatogonia and primary spermatocytes contain 46 chromosomes – 44 autosomes and two sex chromosomes (X and Y).
2. During the second stage, primary spermatocytes divide (meiosis I), forming a secondary generation of spermatocytes, called **secondary spermatocytes.** Each secondary spermatocyte contains 22 autosomes and one sex chromosome (either X or Y).

3. During the third stage, each secondary spermatocyte divides again (meiosis II), forming **spermatids.**
4. During the final stage, spermatids undergo a couple of structural changes which transform them into **spermatozoa** (sperm). Each spermatozoa consists of a head (which contains the nucleus), a neck, midsection, and tail. The tail contains ATP which provides the energy for the movement and swimming of sperm – also called **sperm motility**.

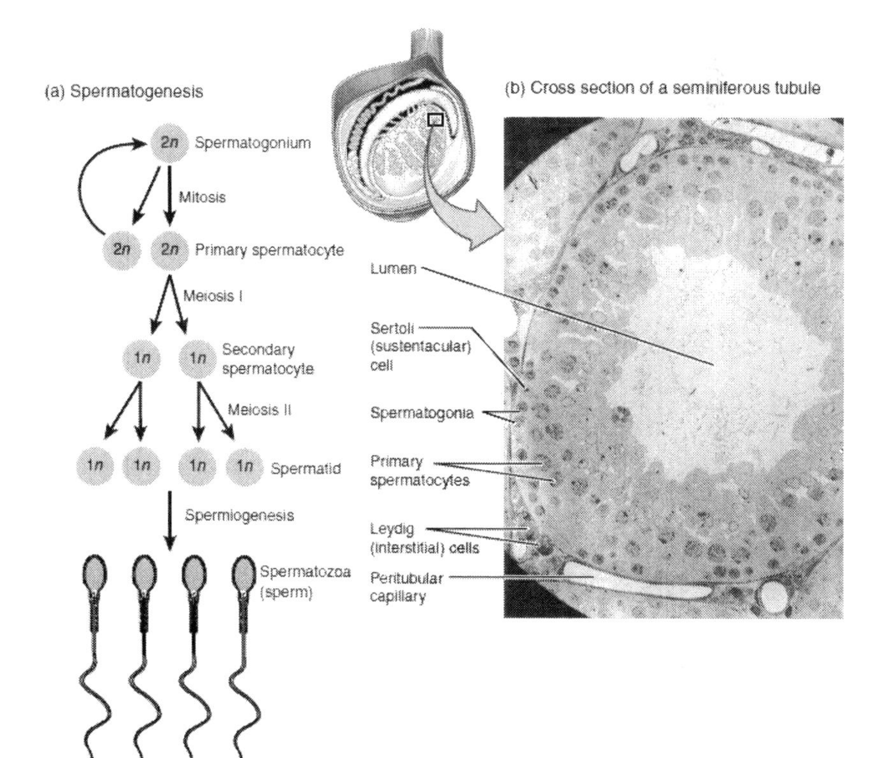

Hormones and the Male Reproductive System

Androgens are male sex hormones which are produced in the testes and the adrenal glands. The main androgens include:

testosterone, follicle-stimulating hormone (FSH), and luteinizing hormone (LH).

Leydig cells, also called **intestinal cells of Leydig**, are located in the testicles adjacent to the somniferous tubules. Leydig cells are responsible for producing testosterone. Testosterone, which is required for spermatogenesis, promotes the development and maintenance of male sex organs. Furthermore, testosterone is also responsible for secondary sex characteristics, which include vocal cord thickness and growth of facial and chest hair.

Two other hormones, LH and FSH, also play an important role in testosterone production. LH stimulates testosterone production from the Leydig cells. FSH stimulates testicular growth as well as increases the production of androgen-binding proteins which increase the concentration of testosterone in the seminiferous tubules, thereby promoting sperm growth.

LH and FSH are important sex hormones in both males and females. LH and FSH trigger estrogen production in the ovaries and testosterone production in the testes.

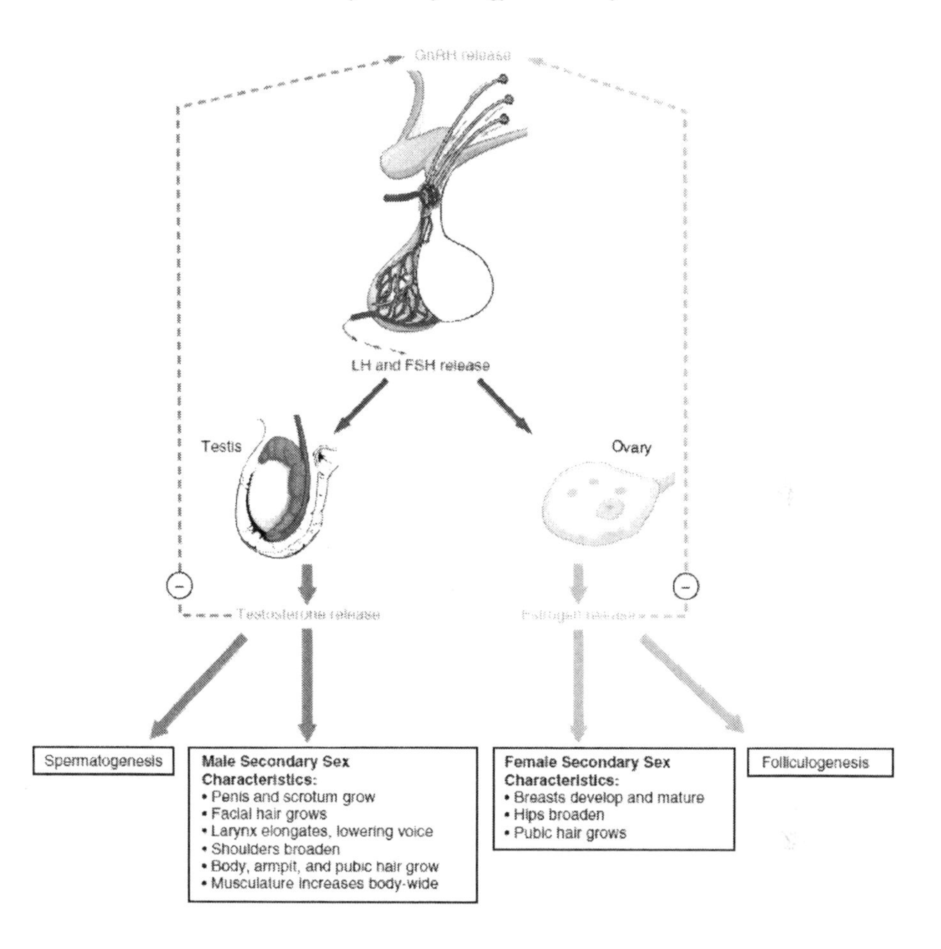
GnRH release
LH and FSH release
Testis
Ovary
Testosterone release
Spermatogenesis
Male Secondary Sex Characteristics:
• Penis and scrotum grow
• Facial hair grows
• Larynx elongates, lowering voice
• Shoulders broaden
• Body, armpit, and pubic hair grow
• Musculature increases body-wide
Female Secondary Sex Characteristics:
• Breasts develop and mature
• Hips broaden
• Pubic hair grows
Folliculogenesis

Section 16: Fluids, Electrolytes, and Acid-Base Balance

The body is an enclosed system which is constantly exposed to a changing external environment. These alterations in the external environment are buffered by responses which help the body maintain its internal environment, which is called homeostasis. Homeostasis it the tendency of the body to seek and maintain a relatively stable equilibrium - a constant optimal internal environment. Both homeostasis as well as the health of the body depends on fluids, electrolytes, and acid-base balance, which allow the body to support its normal physiologic functioning.

Fluid Homeostasis

Body fluids are composed of water and solutes which include amino acids, electrolytes, glucose, and other nutrients. There are four types of fluids in the body:

1. **Extracellular fluid (ECF**), which includes IVF and ISF. ECF is the fluid found in the spaces between cells. ECF accounts for around 20% of human body weight (15% of which comes from ICF and 5% from IVF).
2. **Interstitial fluid (ISF**), sometimes called tissue fluid, is the solution that surrounds tissue cells. ISF fluid accounts for around 15% of human body weight.
3. **Intracellular fluid (ICF**) is the fluid found inside individual cells. ICF accounts for around 40% of a person's body weight.
4. **Intravascular fluid (IVF**), also referred to as plasma, is the extracellular fluid within blood vessels. IVF accounts for around 5% of human body weight.

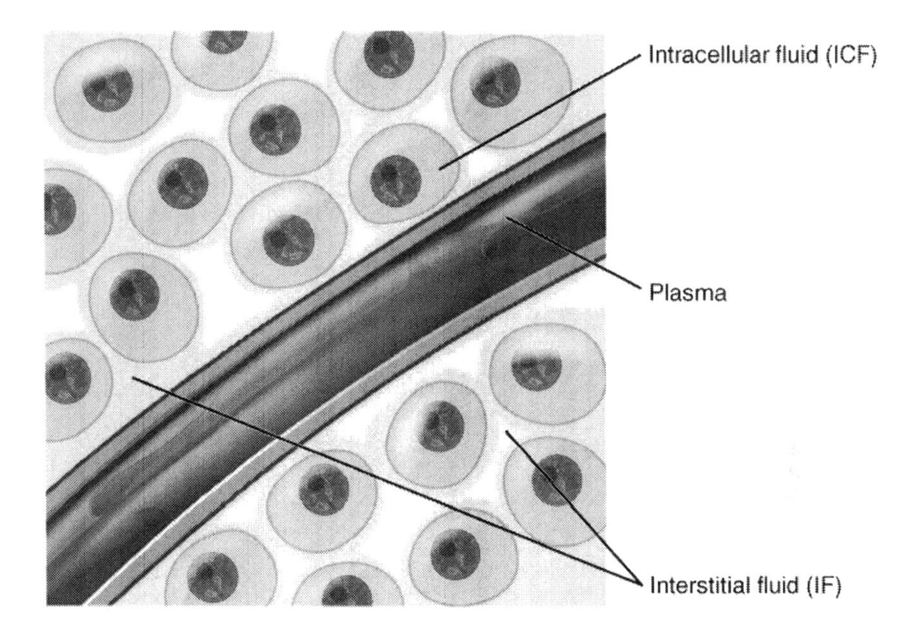

Fluids and solutes constantly move around the body. It is this movement that allows the body to maintain homeostasis. At the cellular level, solutes move through plasma membranes by diffusion, active transport, and osmosis (see ***Section 3***).

- **Diffusion:** Diffusion is a passive transport mechanism by which particles move from an area of high concentration to an area of low concentration area.
- **Active transport:** Active transport is the process by which solutes can move from a low concentration area to a high concentration area. Active transport requires the assistance of a carrier protein and energy supplied by ATP.
- **Osmosis:** Osmosis is the passive movement of fluids across a plasma membrane from an area of lower solute concentration to an area of higher solute concentration.

Water enters the body from the gastrointestinal tract from liquids, foods, and from water of oxidation. Water exits the body through perspiration (skin), expiration (lungs), in stool and in urine. To maintain a healthy body, fluid intake should equal fluid loss. An impairment or abnormality in the mechanisms that regulate fluid balance can result in a fluid imbalance.

Electrolyte Homeostasis

Electrolytes break up into ions when dissolved in water. Ions can be positively charged (**cations**) or negatively charged (**anions**). A balance of cations and anions typically maintains the electrical neutrality of fluids in the body. Examples of cations include calcium, magnesium, potassium and sodium. Examples of anions include chloride, bicarbonate, and phosphate.

Inside the body, electrolytes affect water distribution, osmolarity (the amount of solute per unit volume), and acid-base balance. Several mechanisms help the body regulate and maintain its electrolyte balance:

- **The endocrine system:** The endocrine system helps to keep sodium and potassium levels within the normal range by producing antidiuretic hormones.
- **The gastrointestinal system:** The gastrointestinal system helps regulate gastric juices in the small bowl and in the stomach.
- **The vascular system:** The heart drives the transport of electrolytes in the bloodstream.
- **The kidneys:** The kidneys help regulate electrolytes through its glomeruli filter by filtering out potassium and sodium (smaller particles) and by retaining protein (a larger particle).

The following are the mechanisms which regulate the main

electrolytes contained in body fluid:

- **Bicarbonate:** Bicarbonate levels are regulated by the kidneys.
- **Calcium:** Parathyroid hormones are the main regulator of calcium. Calcium is typically ingested through the GI tract and excreted by the kidneys.
- **Chloride:** Chloride, which moves alongside sodium inside the body, is regulated by the kidneys.
- **Magnesium:** Aldosterone, which controls renal magnesium reabsorption, is the main regulator of magnesium. Magnesium is ingested through the GI tract and excreted in saliva, urine, and breast milk.
- **Phosphate:** Phosphate levels are regulated by parathyroid hormones and by the kidneys.
- **Potassium:** Potassium levels are regulated by aldosterone and by the kidneys. Potassium is mostly absorbed through food and excreted in urine.
- **Sodium:** The kidneys and aldosterone hormones are the primary regulators of sodium. Sodium is absorbed by the body through food and excreted by the skin and the kidneys.

Acid-Base Homeostasis

Acid-base balance is a part of biological homeostasis which is concerned with the proper balance between acids and bases, also referred to as body pH. Whether a fluid is acidic or basic (alkaline), is dependent on the hydrogen ion concentration (H+) of the fluid. pH is a logarithmic measure of hydrogen ion concentration. Water is neutral at a pH of 7.0. The ideal blood pH of the body is 7.4 and a normal pH level ranges anywhere between 7.35 and 7.45 (very slightly alkaline). Acid-base balance is the mechanism employed by the body to keep fluids as close to a neutral pH as possible.

The body maintains its pH level by keeping the ratio of bicarbonate hydrogen ions (HCO_3^-) to carbonic acid (H_2CO_3) at a 20:1 ratio. The body has two natural buffer systems which help regulate blood pH, keeping it within the narrow optimal homeostatic range of 7.35 to 7.45:

- **The lungs:** Respiration plays a crucial role in controlling pH. The lungs retain and excrete carbonic acid in the form of carbon dioxide (CO_2). A decrease in blood pH stimulates respiration. As a result of increased respiration, carbon dioxide (CO_2) in the blood levels decrease which means that less carbonic acid and hydrogen ions remain in the bloodstream. This ultimately leads to an increase in pH.

- **The kidneys:** The kidneys assist in regulating pH in two ways. Firstly, by reabsorbing sodium bicarbonate ($NaHCO_3$) from the urine and secondly, by secreting hydrogen ions into the urine.

Interruption to a buffer system, such as those caused by respiratory or metabolic disorders, can cause an acid-base imbalance:

- **Acidosis** is characterized by an excessively acid condition of tissues and body fluids. Acidosis occurs when there is an accumulation of carbonic acid (H_2CO_3) and a decrease of bicarbonate hydrogen ions (HCO_3^-).
- **Alkalosis** is characterized by an excessively alkaline condition of tissues and body fluids. Alkalosis occurs when there is an increase in bicarbonate hydrogen ions (HCO_3^-) and a loss of carbonic acid (H_2CO_3).

The Buffer Systems in Body Fluids

A buffer is a chemical which binds with either the base or the acid in order to increase or decrease the solution pH. Buffers are produced by cells and made available in the blood. Buffer systems that assist in maintaining acid-base balance include: **carbonic acid-bicarbonate buffer,** the **phosphate buffer,** and **protein buffers.**

The **carbonic acid-bicarbonate buffer system** is the primary buffer system in ECF and is responsible for around 80% of extracellular buffering. This buffer is maintained by the body by eliminating either one of two components: the acid (carbonic acid) or the base (hydrogen carbonate ions). In this way, a strong acid is converted into a weak acid or a strong base is converted into a weak base. The acid can be eliminated through increased respiration. The base can be eliminated through the kidneys.

The **phosphate buffer system** works similar to the carbonic acid-bicarbonate buffer system in that it converts a strong acid into a weak acid or a strong base into a weak base. In the phosphate buffer system, the pH of fluids is regulated as they pass through the kidneys.

The **protein buffer** system can also even out minor fluctuations in pH. In the protein buffer system, proteins such as plasma proteins can release excess hydrogen if needed or absorb hydrogen ions generated by the metabolic process. In this system, pH is therefore controlled by either releasing hydrogen ions in the presence of excess base, or by taking in hydrogen ions in the presence of excess acid.

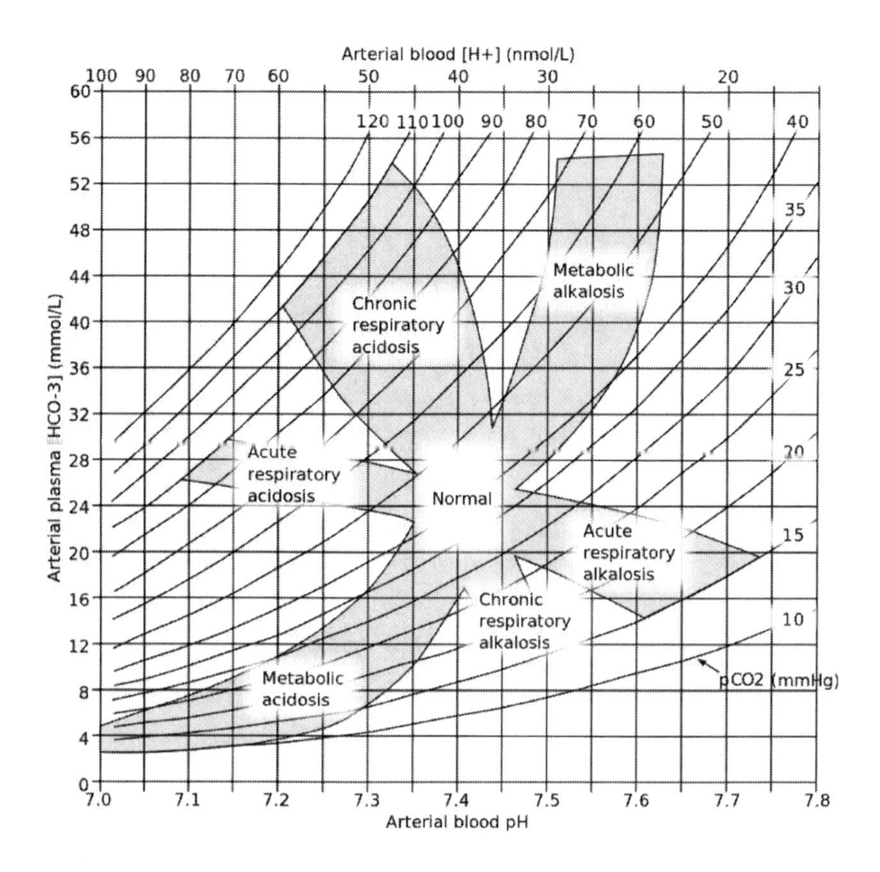
Arterial blood [H+] (nmol/L)
100 90 80 70 60 50 40 30 20
120 110 100 90 80 70 60 50 40
35
30
25
20
15
10
pCO2 (mmHg)
Chronic respiratory acidosis
Metabolic alkalosis
Acute respiratory acidosis
Normal
Acute respiratory alkalosis
Chronic respiratory alkalosis
Metabolic acidosis
Arterial plasma [HCO-3] (mmol/L)
60 56 52 48 44 40 36 32 28 24 20 16 12 8 4 0
7.0 7.1 7.2 7.3 7.4 7.5 7.6 7.7 7.8
Arterial blood pH

Section 17: Nutrition and Metabolism

Nutrition refers to the intake and utilization of nutrients necessary for growth and health. Nutrients are the most basic molecules and atoms needed for survival. Metabolism refers to the chemical processes which occur inside the body of living organisms in order to maintain life. It is through metabolism that the nutrients which are ingested through food are transformed into components which can be used by the body.

Metabolism involves two processes: **anabolism**, also called constructive metabolism, which is the synthesis of simple molecules into complex molecules, and **catabolism**, also called destructive metabolism, which is the breakdown of complex molecules to form simpler ones.

The body requires water and nutrients for growth and health. Most nutrients come from digested food and include the following three main types: **carbohydrates, proteins,** and **lipids.** Besides nutrients, the body also requires **vitamins** and **minerals**. Vitamins promote enzyme reactions and stimulate the metabolism of nutrients whereas minerals are needed for enzyme metabolism, amongst other things.

Carbohydrates

Carbohydrates are organic compounds which contain carbon, hydrogen, and oxygen. They typically include sugars, starches, and cellulose.

Carbohydrates can be categorized as **simple carbohydrates** and

complex carbohydrates.

- Simple carbohydrates, which can raise blood glucose levels rapidly, include fruits, vegetables, and dairy products.
- Complex carbohydrates, which raise blood glucose levels more slowly, include starches and fibers.

Sugars are carbohydrates and provide the primary source of energy for the body. Sugars are categorized as **monosaccharides, disaccharides,** and **polysaccharides.**

- Monosaccharides are simple sugars (e.g. glucose) which cannot be hydrolyzed further to form a simpler sugar. Based on whether the sugar consists of an aldehyde group or a ketone group, it can be categorized as **polyhydroxy aldehydes** (a CHO group, which contains a carbon atom linked to a hydroxyl group) or **ketones** (a CO group, which contains a carbonyl group).

- Disaccharides includes any class of sugar that is synthesized from two monosaccharides. Disaccharides are composed of two monosaccharides minus a water molecule. Examples of disaccharides include lactose (which comprises one glucose molecule and one galactose molecule), maltose (which consists of two glucose molecules) and fructose (which comprises one glucose molecule and one fructose molecule).

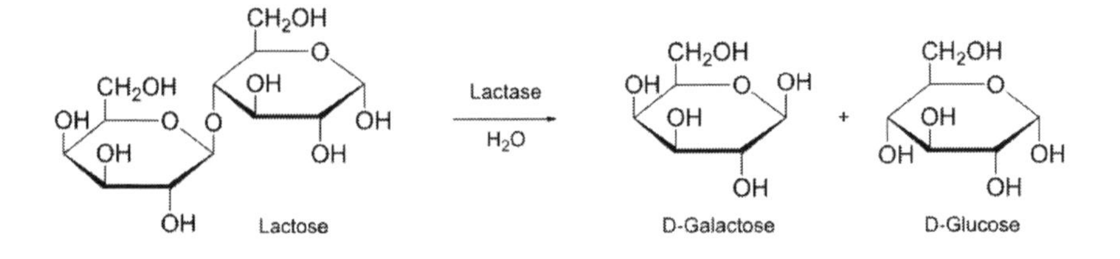

- Polysaccharides are likewise synthesized from monosaccharides. A polysaccharide is composed of a **polymer**, that is, a long chain of 10 or more monosaccharides bonded together by glycoside bonds. Examples of polysaccharides are glycogen and fiber. Fiber can't be broken down into simpler sugars to be used for fuel for the body, whereas glycogen can. The body can also build glycogen from excess monosaccharides (simple sugars) to be used for future use in muscle and liver tissues and as fat in adipose tissue.

Proteins

Proteins are nitrogenous organic compounds which contain large molecules **of amino acid chains**. Some proteins also contain sulfur and phosphorus. In the body, protein is primarily used for growth as well as for the repair of body tissues, but can also be used for energy. Proteins can combine with carbohydrates to form **glycoproteins** or with lipids to form **lipoproteins.**

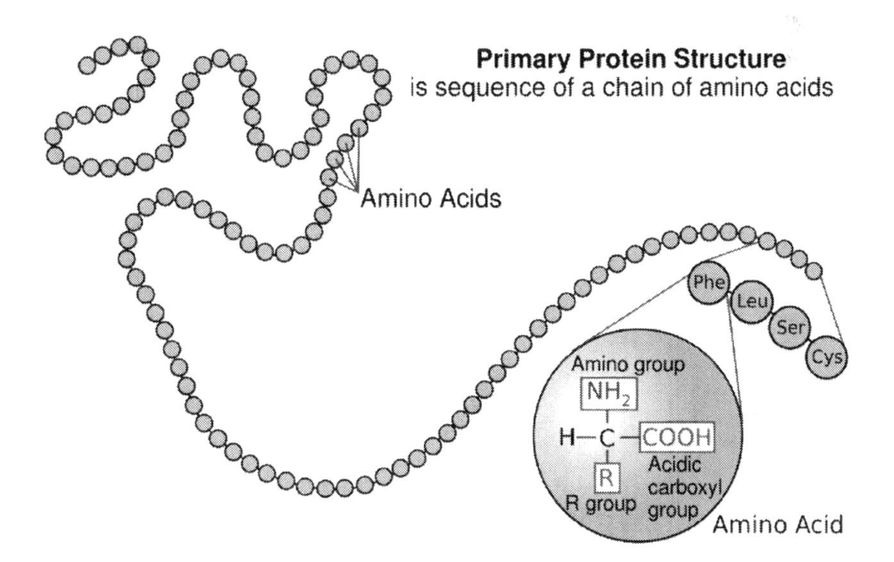

Amino acids, which consist of a carboxyl group and an amino

group, are the basic structural units of proteins. They combine in a process called **condensation,** which is a chemical reaction whereby the carboxyl (COOH) group of one amino acid binds with the amino group (NH2) of another amino acid. The condensation reaction releases a water molecule and forms a **peptide bond** between the amino acids.

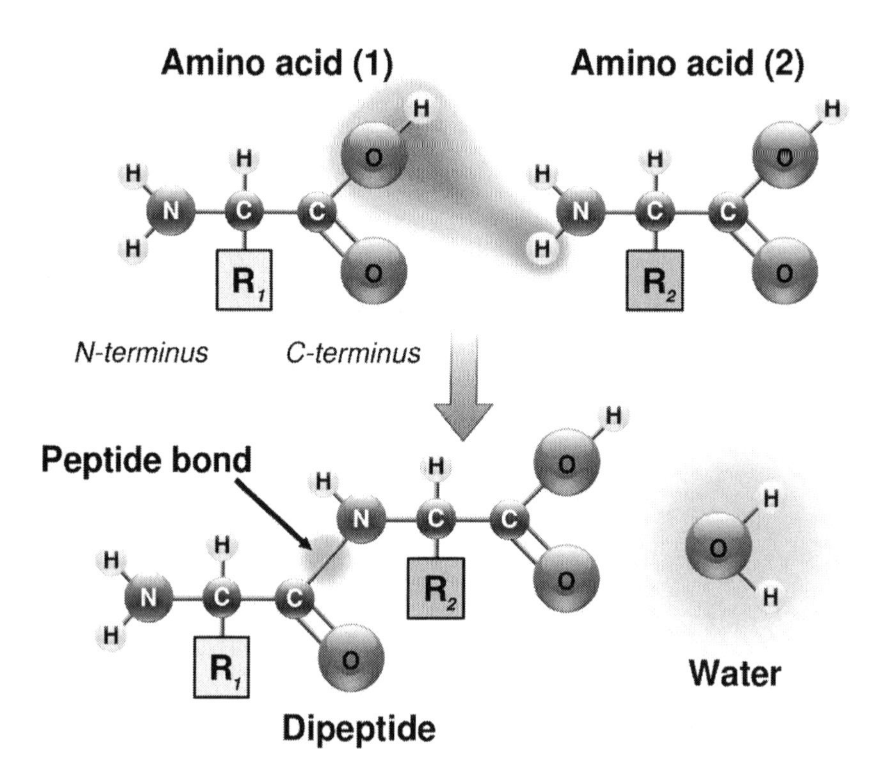

Lipids

Lipids are organic compounds which are insoluble in water but soluble in organic solvents. The main lipids include **fats, phospholipids**, and **steroids**.

- A fat, or triglyceride, is formed from one glycerol molecule and three molecules of fatty acid. A fatty acid is composed of a hydrocarbon chain and a carboxyl group. Fatty acid chains can vary in length.

- Phospholipids are similar to fats but contain one phosphate group as part of their structure. Phospholipids are the main lipids in cell membranes.
- Steroids are formed from four rings of carbon atoms which are attached to various side chains. Steroids contain no glycerol or fatty acid molecules. Common steroids include cholesterol, sex hormones, and bile salts.

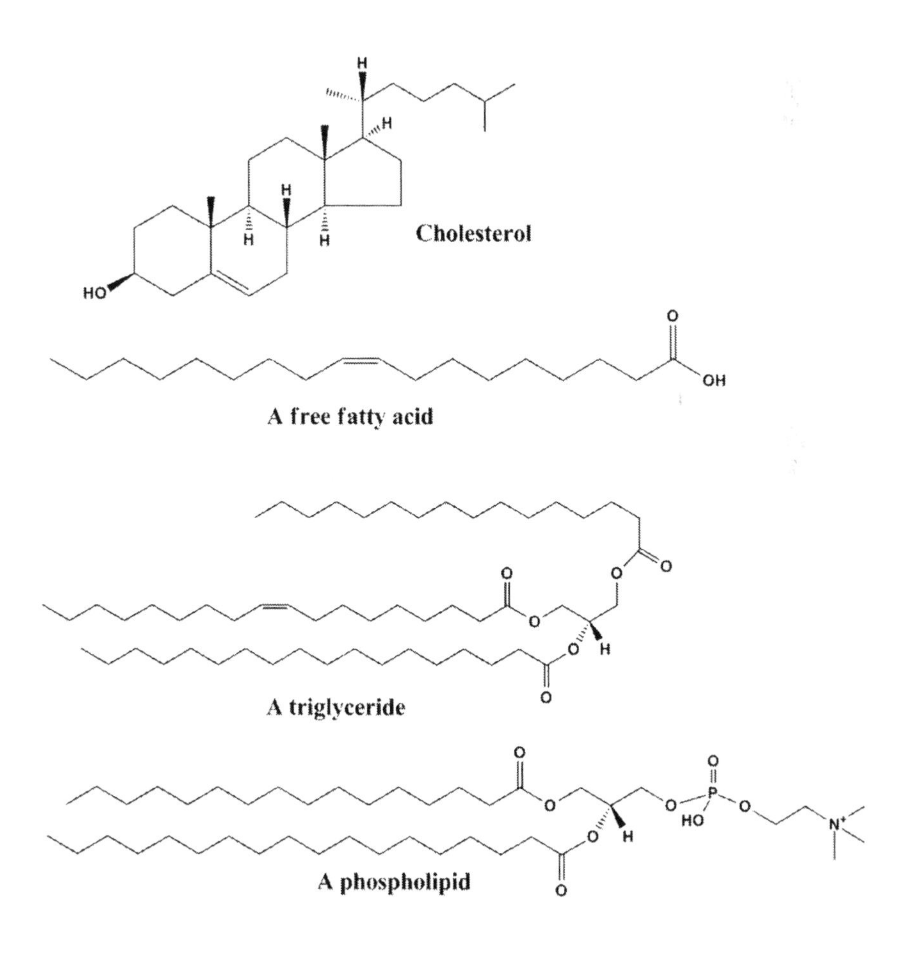

Vitamins and Minerals

Vitamins are organic compounds, required in small quantities by the body, which are essential for growth and nutrition. Vitamins are classified as **water-soluble** or **fat-soluble.**

- Water-soluble vitamins are carried in the body's tissue but cannot be stored in the body. Because of this, water-soluble vitamins need replacing on a daily basis. They are found in plant and animal foods and include B complex and C vitamins.
- Fat-soluble vitamins are dissolved in fat before being absorbed by the bloodstream. Fat-soluble vitamins can be stored in body tissues and in the liver and thus need not be taken daily. Vitamin A, D, E, and K are examples of fat-soluble vitamins.

Water-soluble vitamins

Vitamin	Main Functions	Major Sources
Vitamin B1 (Thiamine)	Helps with energy metabolism, circulation, digestion, growth, muscle tone maintenance, and is also important to nerve function	Meat, fish, poultry, whole-grain or enriched breads, cereals, legumes, dried beans, nuts and seeds
Vitamin B2 (Riboflavin)	Helps with energy metabolism, RBC formation, and cell respiration. Also important for normal vision and skin health	Meat, fish, poultry, milk, milk products, leafy green vegetables, and whole-grain or enriched grains
Vitamin B3 (Niacin)	Helps with energy metabolism and is also important for the nervous system, digestive system and skin health	Lean meat, poultry, fish, whole-grain or enriched breads and cereals, peanut butter, vegetables (especially mushrooms, asparagus and leafy green vegetables)
Vitamin B6 (Pyriodoxine)	Helps with protein metabolism, antibody	Meat, fish, poultry, vegetables, fruits, wheat germ and

	formation, hemoglobin formation, DNA and RNA synthesis, and also contributes to CNS maintenance	**whole grains**
Vitamin B12 (Cyanocobalamin)	**Helps with RBC formation, cellular metabolism, nutrient metabolism, and is important to nerve function**	**Meat, poultry, fish, seafood, eggs, milk and milk products**
Vitamin C (Ascorbic acid)	**Helps with collagen production, RBC formation, iron absorption, healing, and also aids infection resistance**	**Found only in fresh fruits and vegetables, especially citrus fruits and raw dark leafy vegetables**
Folic Acid	**Helps with protein metabolism, RBC formation, cell growth and reproduction, DNA and RNA formation**	**Leafy green vegetables and legumes, citrus foods, milk products and whole grains**

Fat-soluble vitamins

Vitamin	Main Functions	Major Sources
Vitamin A	Helps with body tissue repair and maintenance, vision, bone growth, infection resistance, cell membrane metabolism, and immune system health	Meat, milk, milk products, leafy green and yellow vegetables
Vitamin D (Calciferol)	Helps with calcium and phosphorus metabolism, heart function and nervous system maintenance	Egg yolks, organ meats, fatty fish, liver oils
Vitamin E (Tocopherol)	Helps with cell maintenance, particularly cell membrane stabilization and the maintenance of a healthy immune system	Egg yolks, organ meats, leafy vegetables and wheat germ oil
Vitamin K (Menadione)	Helps with liver synthesis of prothrombin and is also essential for proper blood	Leafy green vegetables

	clotting	

Minerals are solid inorganic substances which are found in the bones, teeth, thyroxin, hemoglobin, and the organs. Like vitamins, minerals are needed to support normal body functions, including enzyme metabolism, growth, muscle contractility, osmotic pressure, membrane transfer of essential molecules, acid-base homeostasis, and nerve impulse transmissions. Minerals are classified as **major minerals** (of which the body needs large amounts of) and **trace minerals** (of which the body needs small amounts of).

- Major minerals include calcium, chloride, magnesium, phosphorus, potassium, and sodium.
- Trace minerals include chromium, copper, fluorine, iodine, iron, manganese, molybdenum, selenium, and zinc.

Major minerals

Mineral	Main Functions	Major Sources
Calcium	Important for proper blood clotting, bone and tooth formation, muscle growth and contraction, and also helps with nerve functioning, blood pressure regulation and immune system health	Milk, milk products, grains, nuts, legumes and leafy vegetables
Chloride	Helps with the maintenance of fluid, electrolyte, acid-base balance and osmotic pressure	Table salt, fruits and vegetables
Magnesium	Needed for protein synthesis, cellular respiration, muscle contraction, nerve impulse transmission and also supports immune system health	Nuts and seeds, cocoa, legumes and leafy green vegetables
Phosphorus	Helps with bone and teeth formation, energy production, cell growth and cell repair	Eggs, fish, meat, fish, poultry, milk and milk products
Potassium	Helps with muscle contraction, proper fluid balance, nerve transmission, fluid distribution, osmotic pressure balance and	Fresh fruits, vegetables, nuts and seafood

	acid-base balance	
Sodium	**Needed for proper cellular fluid maintenance, acid-base balance, nerve transmission and muscle contraction**	**Table salt, soy sauce, milk, cheese and processed foods**

Trace minerals

Mineral	Main Function	Major Sources
Chromium	Helps regulate blood glucose levels	Liver, whole grains, nuts and cheese
Copper	Helps with hemoglobin formulation, iron metabolism, and plays a role in respiration	Legumes, nuts and seeds, whole grains, organ meats and drinking water
Fluoride	Helps with the bone and teeth formation of bones and helps prevent tooth decay	Drinking water and seafood
Iodine	Iodine is found in thyroid hormones and helps with growth, development, energy production and metabolism	Iodized salt, seafood and foods grown in iodine-rich soil
Iron	Helps with RBC formation, growth (in children), cellular respiration, and also aids stress and disease resistance	Egg yolks, organ meats, fish, poultry, oysters and green vegetables
Manganese	Helps form connective tissue, bones, sex hormones, and also aids blood clotting	Found especially plant foods

Molybdenum	**Functions as a cofactor for enzymes**	**Leafy green vegetables, legumes, milk, bread and grains**
Selenium	**Helps with cellular protection, immune mechanisms and the synthesis of mitochondrial ATP**	**Meats, seafood and grains**
Zinc	**Needed for protein and DNA synthesis, burn and wound healing, metabolism, carbohydrate metabolism, cell growth, organ growth and development**	**Liver, meats, seafood, whole grains and vegetables (particularly mushrooms)**

Nutrient Digestion, Absorption and Metabolism

Nutrients are digested in the GI tract in a process called hydrolysis. Hydrolysis is the chemical decomposition whereby a compound unites (reacts) with a water molecule and splits into simpler compounds.

Carbohydrate Digestion, Absorption and Metabolism

Enzymes are responsible for breaking down complex carbohydrates into simpler compounds. In the mouth, **salivary amylase** starts the process hydrolysis (from a polysaccharide into a disaccharide). This process is then continued in the small intestine by an enzyme called **pancreatic amylase.**

Different enzymes are responsible for splitting (hydrolyzing) different disaccharides into monosaccharides. **Lactase** for example hydrolyzes lactose into glucose and sucrose and **sucrase** hydrolyzes sucrose into glucose and fructose.

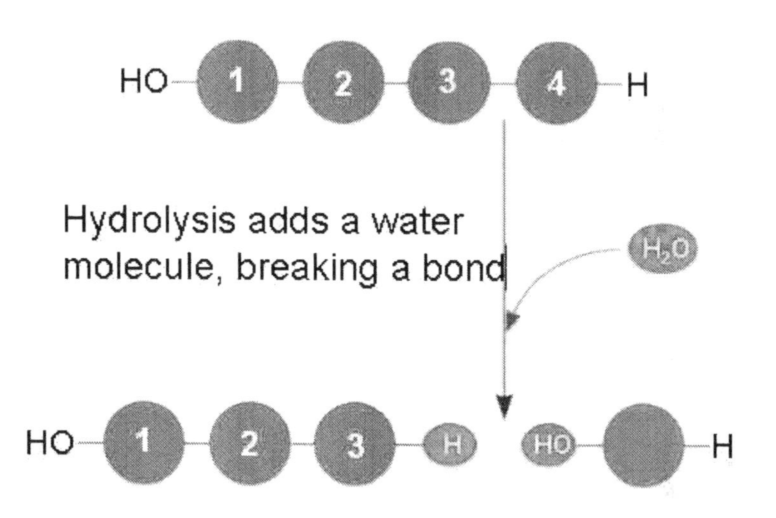

Once the compound is broken down into a monosaccharide, it is then absorbed by the intestinal mucosa and transported into the liver. Inside the liver, enzymes convert galactose and fructose into glucose. The glucose is then absorbed by the intestinal mucosa.

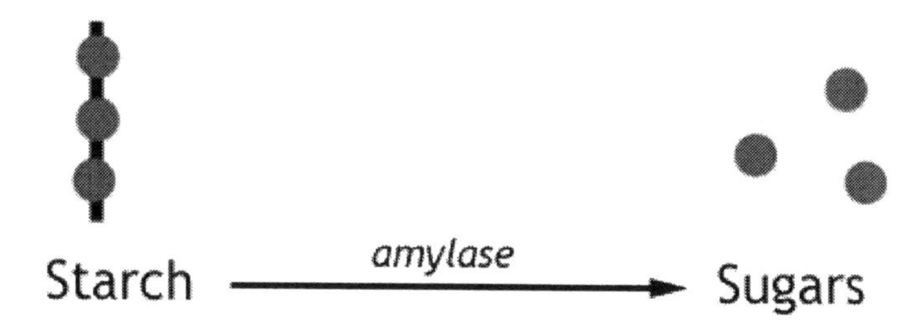

Glucose is the main source of energy in the human body. **Carbohydrate catabolism** is the breakdown of carbohydrates into ATP, which is achieved through the oxidation of glucose molecules. Energy from **glucose oxidation** is generated in three phases: **glycolysis,** the **Krebs cycle (also called citric acid cycle),** and the **electron transport system.**

1. **Glycolysis:**

Glycolysis is the first step in the oxidation process of glucose, which occurs in the cell cytoplasm. Glycolysis yields energy in the form of ATP and acetyl CoA.

During glycolysis, the (6-carbon) glucose molecule is broken down into two molecules (3-carbon) **pyruvic acid**, also called **pyruvate**. Pyruvate is an organic acid which supplies cells with energy. Glycolysis also releases energy in the form of **ATP**.

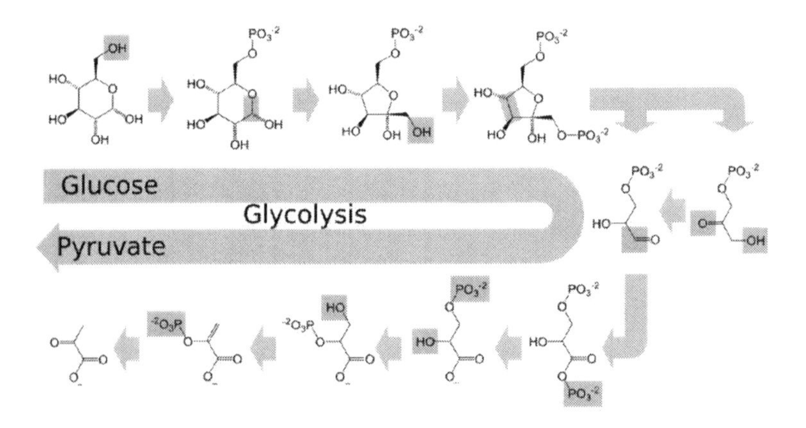

The pyruvate molecules produced by glycolysis also release a carbon dioxide (CO2) molecule. Inside the mitochondria, this CO2 molecule is converted into a (2-carbon) acetyl fragment which combines with a coenzyme A (CoA) to form **acetyl CoA.** The molecule acetyl CoA is an important molecule in metabolism. This conversion process also produces one NADH.

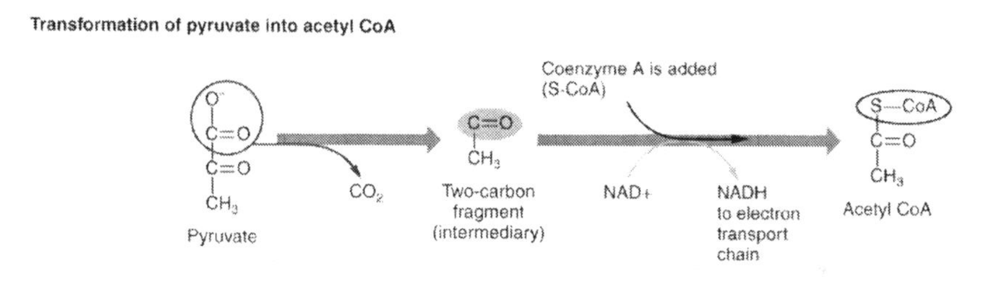

2. **Krebs Cycle:**

The Krebs cycle, also called the citric acid cycle, is the second phase in glucose oxidation. In this process, acetyl CoA is oxidized by enzymes in order to yield energy.

During the Krebs cycle, (2-carbon) **acetyl fragments of acetyl CoA** bind with (4-carbon) **oxaloacetic acid**, forming (6-carbon) **citric acid.** Then CoA molecule from the acetyl CoA separates from the acetyl group in order to bind with more acetyl molecules to form acetyl CoA. Enzymes convert citric acid into different intermediate compounds and back into oxalocetic acid.

This process, which generates energy in the form of ATP, releases CO2 molecules. The Krebs cycle also releases hydrogen atoms which are then picked up by coenzymes NAD (nicotinamide adenine dinucleotide) and FAD (flavin adenine dinucleotide). Each cycle results in three NADH and one FADH2.

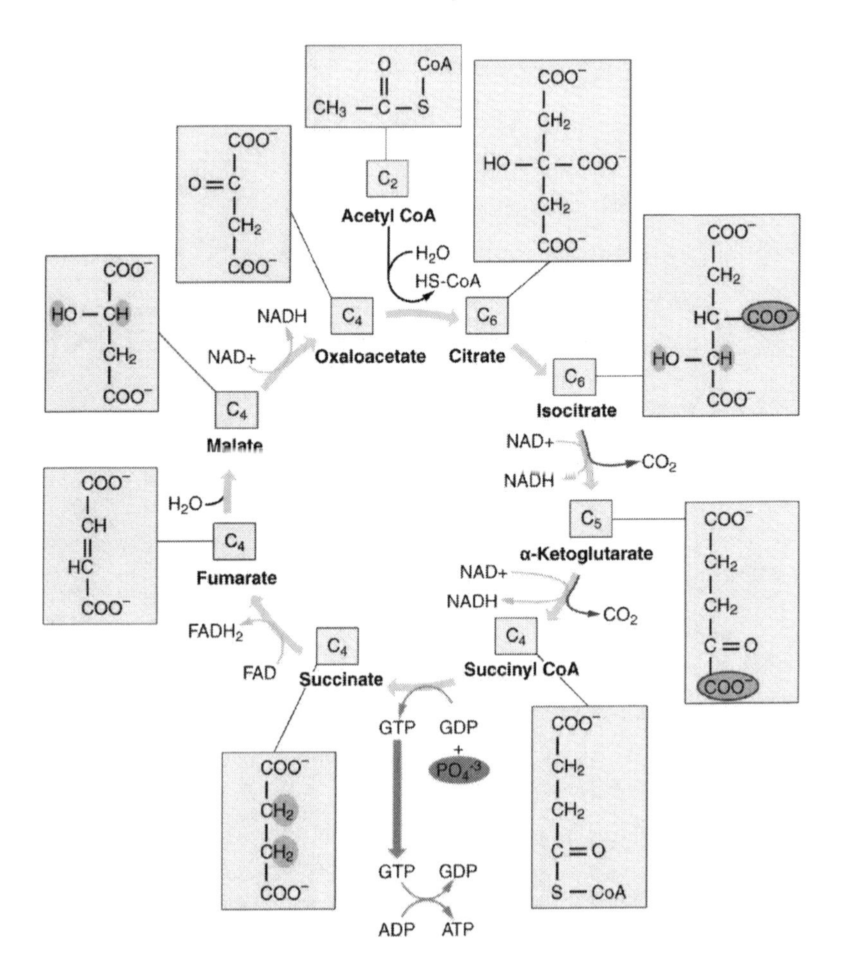

3. Electron Transport System:

The third phase of glucose oxidation, which also takes place inside the mitochondria, is the electron transport system. During this phase, carrier molecules in the inner mitochondrial membrane capture the hydrogen atoms which are carried by NADH and FADH2. Each hydrogen atom carries one hydrogen ion and one electron. The electrons are then passed along the transport system, where they undergo a series of **oxidation-reduction reactions** and release energy as they move through.

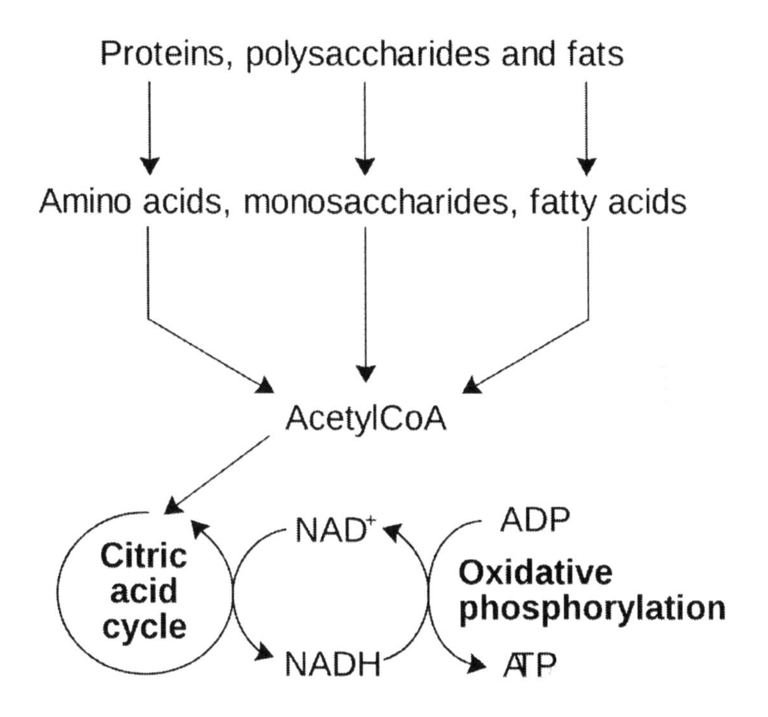

The released energy is then used to transport positively charged ions back and forth across the membrane which separate the two parts of the mitochondria (the intermembrane space and the matrix of the mitochondrion). The energy generated from this movement is stored in ATP.

For this final step, oxygen is required. Oxygen attracts electrons along the chain of carriers in the electron transport system. Because of this, the process is called **oxidative phosphorylation.** After passing through the electron transport system, the hydrogen ions bind with oxygen to produce water.

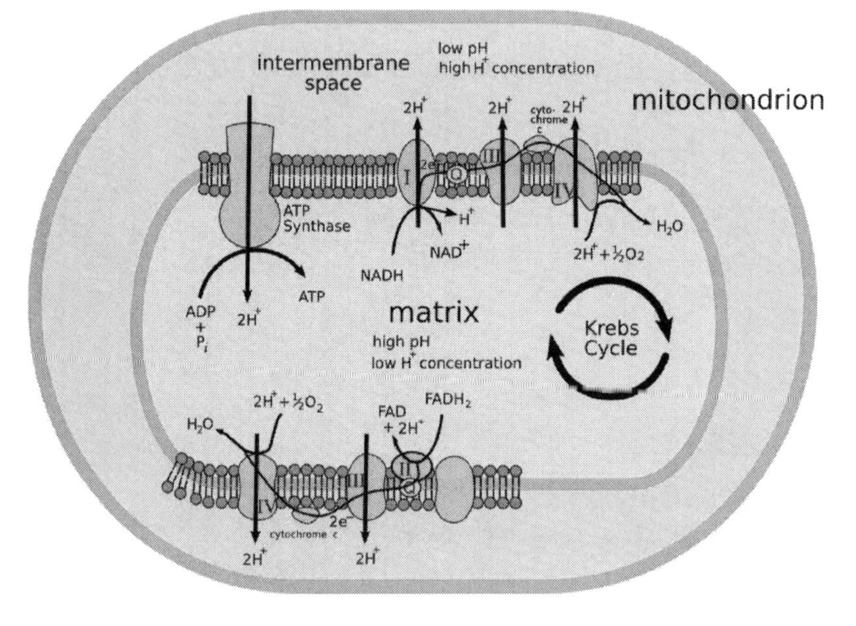

All carbohydrates ingested by the body are converted into glucose. Any glucose not needed for immediate energy is either converted into lipids or stored by the body as glycogen. The liver, muscle cells and certain hormones play a crucial role in controlling and regulating blood glucose levels.

The liver plays a crucial role in regulating blood glucose levels:

1. When there is too much glucose in the body, hormones will stimulate the liver to convert glucose into lipids or glycogen.

- **Lipogenesis:** Glucose can be converted into lipids in a process called lipogenesis.
- **Glycogenesis:** Glucose can be converted into glycogen in a process called glycogenesis.

2. If there is a shortage of glucose in the body, the liver can form glucose through two processes:

- **Glycogenolysis:** The liver can form glucose by breaking glycogen into glucose through a process called glycogenolysis.
- **Gluconeogenesis:** Alternatively, the liver can also synthesize glucose from amino acids through a process called gluconeogenesis.

Hormones can stimulate specific metabolic processes within the body. With regards to blood glucose levels, insulin is the only hormone which can significantly reduce blood glucose levels in the body. Insulin encourages cell uptake and use of glucose for energy, and also stimulates lipogenesis and glycogenesis.

Like the liver, muscle cells can also convert glucose into glycogen. Muscles cells however, don't have the enzymes needed to convert glycogen back into glucose. During extended exercise, muscle cells break down glycogen in a process which produces lactic acid and energy. This leads to a lactic acid build-up in the muscles as glycogen stores in muscle cells become depleted. The build-up of lactate is what results in the burning sensation felt in active muscles after vigorous exercise.

Lactic acid can be used by the body in two possible ways. Some of the accumulated lactic acid in the muscle cells is converted into pyruvic acid which is then oxidized again by the Krebs cycle and the electron transport system in order to generate energy. Some of the lactic acid travels to the liver where it is converted to glycogen. The liver converts this glycogen into glucose. The newly formed glucose then travels back through the blood stream to the muscles where it is stored again as glycogen.

Protein Digestion, Absorption, and Metabolism

Proteins digestion occurs by hydrolysis of the peptide bonds that bind together the amino acids. The protein digestive enzyme is called **protease.** After hydrolysis, the amino acids are absorbed by the intestinal mucosa from where they are transported to the liver.

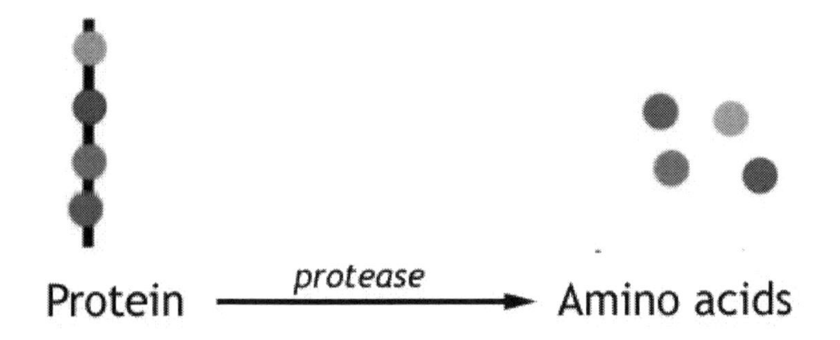

Absorbed amino acids mix with other amino acids in the **amino acid pool.** The human body cannot store amino acids. Because of this, amino acids are converted into protein, glucose, or straight into energy. This conversion however, first necessitates transformation of the amino acid by transamination or deamination.

- **Transamination:** Transamination is the process by which one amino group is exchanged for a keto group in a keto acid. This process is facilitated by **transaminase** enzymes and results in the formation of a new amino acid and a new keto acid.
- **Deamination:** Deamination is the the process by which one amino group is removed from an amino acid, forming one ammonia molecule and one keto acid. Most of the ammonia is then converted to urea and excreted from the body through urine.

Amino Acid Metabolism

Amino acid synthesis is the process by which various amino acids are produced from other compounds. Amino acids are classified as **essential** and **nonessential.** Essential amino acids are amino acids which cannot be synthesized by the body. They are essential in the sense that they must be ingested through food. Nonessential amino acids are amino acids which are not essential in the human diet as they can be synthesized by the body.

Essential amino acids include:

- Histidine
- Isoleucine
- Leucine
- Lysine
- Methionine
- Phenylalanine
- Threonine
- Tryptophan
- Valine

Nonessential amino acids include:

- Alanine
- Arginine
- Asparagine
- Aspartic acid
- Cystine
- Glutamine
- Glycine
- Hydroxyproline
- Proline
- Serine
- Tyrosine

Proteins are synthesized from 20 amino acids: 9 essential ones and 11 nonessential ones. All amino acids not needed for protein synthesis are converted into glucose by the liver. Depending on their characteristics, amino acids which are not used to form proteins are either metabolized by the Krebs cycle and the electron transport system to generate energy or converted into keto acids.

- **Gucogenic amino acids** are the main carbon source for gluconeogenesis. They can be converted into glucose through gluconeogenesis. Glucogenic amino acids can either be catabolized for energy or converted into glycogen or fatty acids for energy storage.
- **Ketogenic amino acids** can be used for lipid synthesis or ketogenesis. Ketogenic amino acids don't have the ability to form glucose.
- **Some amino acids are both glucogenic and ketogenic.**

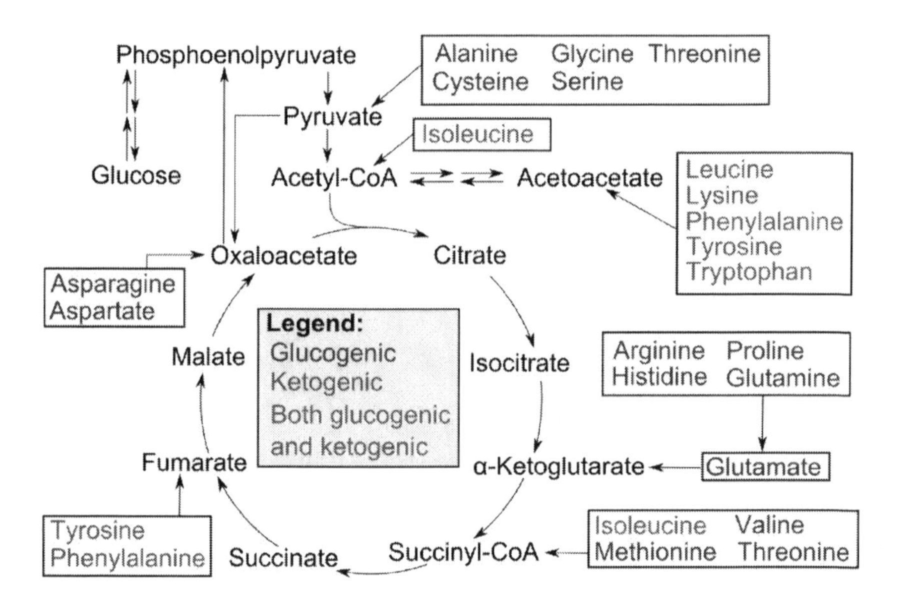

Lipid Digestion, Absorption and Metabolism

Lipid digestion occurs primarily in the small intestine. The lipid digestive enzyme is called **lipase.** Lipase hydrolyzes the bonds between glycerol and fatty acids in a process which restores the water molecule, and which breaks phospholipids down into glycerol, short-chain fatty acids, long-chain fatty acids, and monoglycerides.

Glycerol molecules diffuse directly through the mucosa and short-chain fatty acids diffuse into the intestinal epithelial cells from where they are transported to the liver. Long-chain fatty acids and monoglycerides dissolve in the bile salt micelles before diffusing into the intestinal epithelial cells. Monoglycerides are further broken down into glycerol and fatty acids by the enzyme lipase.

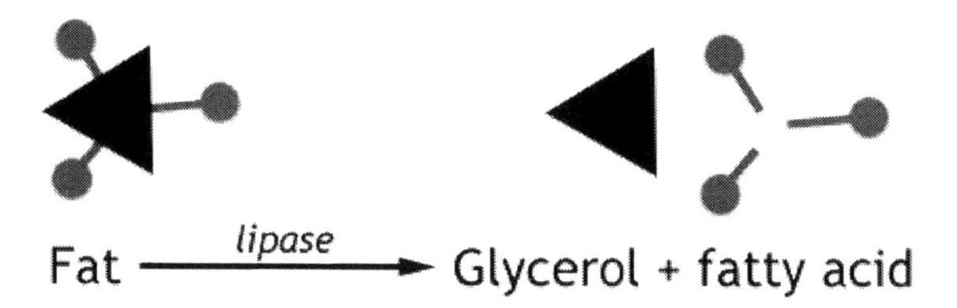

Lipids are stored in adipose tissue. When needed for energy, lipase hydrolyzes each lipid molecule into one glycerol molecule and three molecules of fatty acids. The glycerol can be converted into pyruvic acid which then enters the Krebs cycle. The fatty acids are catabolized by **beta oxidation.** Beta oxidation produces two carbon units which bind with CoA to form **acetyl CoA**.

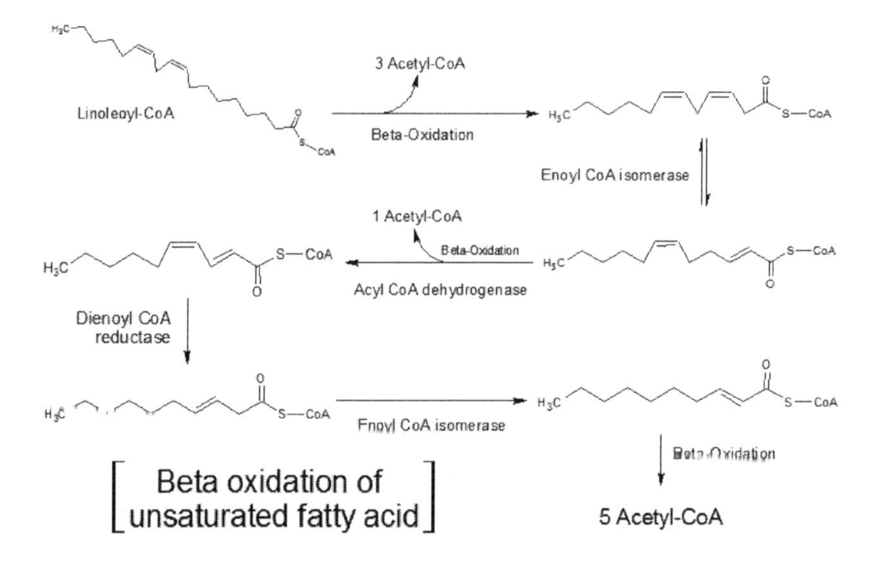

Ketogenesis

Fatty acids are broken down in a process called ketogenesis, which results in the formation of ketone bodies. Ketone bodies are three water-soluble molecules (**aceotoacetate, beta-hydroxybutyrate,** and **acetone**) that are produced by the liver from acetyl which are largely derived from fatty acid catabolism. Ketone bodies can be used for energy by body tissues, including muscle tissue and brain tissue.

Under conditions such as fasting, starvation, or uncontrolled diabetes, the body produces more ketone bodies than it can use for energy. In such conditions, the body uses fat rather than glucose as its primary energy source. When the body breaks down fats for energy, which creates fatty acids through beta oxidation, ketones are created. This in turn leads to an excess of ketone bodies which disturbs the body's acid-base balance – a state called ketosis.

Section 18: Final Notes

"It's more important to understand the imbalances in your body's basic systems and restore balance, rather than name the disease and match the pill to the ill."

-***Mark Hyman MD***

I would like to take this opportunity to thank you for purchasing this book. I hope you now have a solid foundation of the fundamental structures and functions of the human body. Of course, there is a lot more to the human body and I do encourage everyone to continue their exploration, as there is always more to discover and learn.

I sincerely wish you the best of luck and the best of health!

Best wishes,

Dr. Phillip Vaughn

33405900R00177

Made in the USA
Middletown, DE
12 July 2016